ROYAL ENFIELD THE POSTWAR MODELS

NITON PUBLISHING

ROYAL ENFIELD THE POSTWAR MODELS

The 125, 150, 250, 350, 500, 700, 750 Singles and Twins

Roy Bacon

Published by Niton Publishing
P.O. Box 3, Ventnor, Isle of Wight, PO38 2AS

First published in 1982 in Great Britain by Osprey Publishing Limited,
27A Floral Street, London WC2E 9DP
Member company of the George Philip Group

A CIP catalogue record for this book is available from the British Library

ISBN 1 85579 019 X

Original edition:-
Editor Tim Parker
Design Gwyn Lewis
Reprinted through World Print Ltd.

Contents

Acknowledgements 6
Foreword by Johnnie Brittain 7
1 Ancestry 8
2 Big Single 16
3 Bullet 30
The American Indian connection 45
4 Twin 50
5 Big Twin 64
6 Interceptor 78
7 Clipper, Crusader, Turbo 90
8 GP5 122
9 Flying Flea 130
10 Indian Enfield 146
11 Competition 152

Appendix
1 Specifications 164
2 Colours 176
3 Numbering system 180
4 Carburettor settings 181
5 Prices 183
6 Model recognition points 186
7 Model charts 191

Acknowledgements

The first motorcycle I rode, the first I owned, and my first new machine were all Enfields of one sort or another which must prove something, even if only that they were available. My first bike was a 1936 225 cc two-stroke and it carried me for many happy miles.

So writing this history was especially interesting and, as always, much assisted by other people to whom I owe my thanks. Many of these are members of the owners club and include Steve Mayhew, Don McKeand, Peter Lovett and Ivor Mutton. To Ivor especially goes my thanks for the use of so many of his unique collection of photographs of the marque and for entertaining me for a day while I looked at them and tried to select from the many. Seldom has an author been so spoilt for choice or so kindly looked after.

I must also thank Ted Boyle and Nigel Halliday of the Berkeley Owners Club who supplied information and pictures of that machine, also Mike Jackson of Andover Norton for 175 cc Crusader gen.

The bulk of the photographs used in this book come from Ivor Mutton's collection, and the remainder from magazine files. I am again indebted to Bob Berry and Peter Law of *Motor Cycle News*, Mick Woollett of *Motor Cycle Weekly* and Mike Nicks of *Classic Bike* for their help. Other material came from the National Motor Museum at Beaulieu and many pictures were originally Enfield publicity shots taken by Boswell, Barratt and Phillips.

After picture selection all material was returned to owner and some carried the imprint of a freelance photographer. In all cases I have tried to make contact to clear copyright but if my letter failed to reach you or I have unknowingly used an unmarked print, I can only apologise.

The freelance photographers and agencies whose pictures were used in addition to those mentioned above are: Ian Buckden, Malcolm Carling, P. R. Dicks, Derek Evans, Brian Holder, Industrial Photographic, K. G. Jones, T. C. March, Andrew Morland, and E. A. Winpenny.

Finally I must thank Johnny Brittain for his foreword and Tim Parker the editor for his help in writing this book.

I hope you enjoy it.

Roy Bacon
Hampton, Middlesex
June 1982

This is one more of my Osprey titles now published under my Niton imprint. The text has been altered to bring the story of the Madras-built machines up to date, and an insert added to cover the American Indian connection of 1955-59. The colours used by some of the models built for the USA have been included.

Niton, January 1992

Foreword

Johnny Brittain with the ACU Trials Star he won in 1956. As always, it is the 'drivers' award although you ride a solo, and drive a combination

I feel honoured to have been asked to write the foreword and to be associated with this book.

Of course I'm biased but I have read this entertaining work on the postwar Royal Enfields with great pleasure. I'm not an expert myself on every single machine they made but I can't fault Roy Bacon's analysis—in fact, in spite of my years with the company, there is much I have learned.

Looking back and being reminded of what happened in the trials and ISDT scene in the 1950s has brought back to me many happy memories and I'm amazed at just how successful the Enfields were and, come to that, how the Brittain family seemed to play such a role. My father, Vic, started his career in the late 1920s until 1939 when the war suspended activities. When hostilities ceased he joined Royal Enfield to continue his career. His highlight was in the 1948 ISDT in San Remo where he and Charlie Rogers riding Bullets, won Gold Medals and were in the winning British Trophy team. It was a natural step that I followed in his footsteps for much of the next decade and into the 1960s, and that I was joined by my brother, Pat.

We were part of a 'family company' and we felt privileged and rather special in fact; my ambitions were to be a good trials rider and an ambassador for the Royal Enfield Company. Enfields weren't a large company and they didn't have the cash resources to splash around on anything but I am sure that Frank Smith appreciated my aims and we had our place in history. I believe the developments the works riders undertook improved the road production machines. I would like to pay tribute to all my machines looked after by Charlie Rogers, for in the 15 years I rode for Royal Enfield their reliability was outstanding.

I'm therefore delighted to be able to give this book my seal of approval and suggest that every British bike enthusiast should read it, and every Enfield buff must read it. The excellent photographs alone are worth all those words.

Johnny Brittain
ex Royal Enfield works trials rider
John Brittain (Motor Cycles)
Bloxwich, Walsall, West Midlands.
May 1982

1 Ancestry

The Royal Enfield name can be traced back to mid-Victorian times for the company who built the machines had its roots buried in the middle of the nineteenth century. It was during this period that the firm of George Townsend and Company was founded in the hamlet of Hunt End near Redditch in Worcestershire to make sewing needles and machine parts.

As with all small machining firms, even up to the present day, they filled out their order books with general work while retaining their special skills for the needles, and thus became drawn into the bicycle world during the 1880s. By 1890, when they had become a limited company, they were making their own brand of machine along with many fittings for sale to the trade.

Many changes occurred over the next two years with Albert Eadie becoming managing director and Robert Walker Smith works manager, while the company name changed to Eadie Manufacturing. At this time the Enfield name was first used for a new range of bicycles because of contracts with the Royal Small Arms Factory at Enfield in Middlesex. The machines were shown at premises in Edmund Street, Birmingham, run by Horace Granham and, in 1893, the word 'Royal' was added to their name.

In the same year appeared the famous slogan 'Made like a Gun', along with the field gun trade mark. This was to stay with them through the years, although around the turn of the century they also used as a trade mark a rifle with fixed bayonet impaling the words 'Royal Enfield'. Also in 1893 the Enfield Manufacturing Company Limited was formed to market the bicycles, while the Eadie company continued to make them.

Three years later, in 1896, the New Enfield Cycle Company was formed to embrace both the bicycle making division of Eadie and the sales of Enfield. Within a year the prefix 'New' had been dropped and the Enfield Cycle Company was firmly launched with the name it was to keep throughout its history. The Eadie company itself continued to make more bicycles and, later on, motorcycle parts before being acquired by BSA, and in 1907 Albert Eadie left Enfields for a seat on the BSA board.

This left Robert Smith in charge to be appointed managing director in 1913. Six years later his eldest son, Frank Walker Smith, joined him as assistant managing director and on the father's death in 1933 the son, who had actually been in charge for many years, succeeded him to take over the company. This stability of management continued on for Frank Smith stayed at the helm as managing director and later chairman until his death in 1962, with his brother G. H. Smith also on the board and acting

The 500 cc Bullet which retired in the 1934 TT. Cecil Barrow finished 8th the following year on a similar model

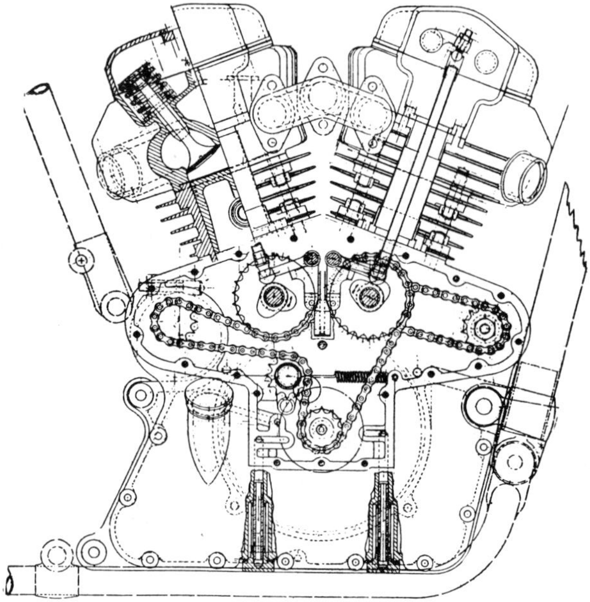

The chain drive timing gear of this 692 cc vee twin prototype of 1937 was used in a simpler form in the postwar twins

as company secretary, while two more brothers also worked in the business.

Their first powered machine was a quadricycle built in 1898 and this type, together with a tricycle, were built using De Dion engines for some years. A quad ran in the famous 1000 Mile Trial of 1900, so starting what was to become a long Enfield tradition. In 1901 the first motorcycle was built with the engine over the front wheel driving the back one with a crossed belt. This was followed by a machine with the engine behind the saddle tube and then, in 1903, with it in the more usual central position, just in front of the bottom bracket and the pedals. It was an advanced design and had what was to become a classic Enfield feature later on, the oil was carried in an extension of the sump. It also had a gear primary reduction to a countershaft and chain drive to the rear wheel.

They soon reverted to the more accepted direct belt drive and for a small range of conventional singles, but did adopt mechanically operated side valves early on. However, their real interests in the Edwardian period were making parts for the trade and in motor cars. The first of these was built as early as 1901 and all this activity increased the size of the Hunt End works and led to a new factory being built at Redditch in 1907, destined to become the company's home.

In 1910 the firm revived its interest in motorcycles and moved into the field of vee twin-engined models. They were in the forefront of the industry in adopting the countershaft gearbox and chain drive which appeared on the 1911 twin and were soon to adopt other advanced ideas for the time. Among these were dry sump lubrication in place of the hand operated pump and enclosed valve gear to help keep the oil inside the engine.

The company expanded and prospered and after the first war let the old Hunt End factory go to the Nife battery concern, while the Redditch plant was extended. Through the twenties the machines were conventional and more than capable of holding their own with most of the parts being made in the Redditch plant, although they also used proprietary engines and Sturmey Archer gearboxes on occasion.

In 1930 engines appeared again with the oil tank cast-in with the crankcase and, as early as 1931, a model with a four valve cylinder head was on offer. The next year a 225 cc two stroke was available, a model first built as early as 1914 but dropped late in the 1920s. There was also a 150 cc version which caused a sensation for it was fully enclosed with built-in legshields.

In 1933 came a 150 cc four stroke with fully enclosed valve gear, another advanced idea at a time when many makers still fitted grease nipples to the rocker spindles, and the first of the Bullet range of sports machines. These were based on the standard models but had improved performance for the sporting rider. At first this was

One of the pre-war works trials bikes complete with high level exhaust and the competition tyres then used

obtained, on the 500, by using the exposed four valves in the penthouse head but in 1935 a fully enclosed three valve design was adopted, the cylinder being inclined.

The following year, in 1936, it was back to the exposed four valves but on an upright cylinder with engine dimensions and timing side layout that were to remain unchanged.

1938 brought the enclosed two valve head back again with the four valves an option for an extra £2 0*s* 0*d* on the 500 cc Bullet but £3 15*s* 0*d* when added to the 500 Competition model. All three sizes of Bullet model could be fitted with the Lucas or BTH racing magneto and this meant a £2 0*s* 0*d* reduction when used in place of the mag-dyno.

The Enfield range was extremely comprehensive in the late 1930s with side and overhead valve machines of 250, 350 and 500 cc, the tiny 150 ohv, the 570 side valve, the two stroke and the 1140 cc side valve, big vee twin. Most models were available in more than one specification with the Bullets and Competition machines to supplement the list.

By 1939 the format of the singles was set and was to change little while they used the magneto to fire the plug. The cylinder sat dead upright on the crankcase and both it and its head were in cast iron, although the 1939 350 cc Bullet did break that tradition by having an all alloy engine. Below the cylinder the crankcase was extended to form the dry sump oil tank in front of the flywheel compartment. The oil was moved by pumps contained in the timing cover with a filter below them, while under the cover lay a gear train from the crankshaft, via the camshafts and on up to the mag-dyno mounted behind the cylinder.

The gearbox was Albion and the rear chain drove a hub containing the Enfield patented cush drive. The cycle parts were standard British industry for the time and the result was a machine that sold well despite the lack of racing involvement. Like BSA, Enfield shunned the hard

The 250 cc model D/D. About 1000 of these side valve machines were made to be used mainly for army training

The wartime model C, 350 cc side valve, used by the services together with the very similar ohv model CO

stuff but were very active and successful in all manner of off-road competitions with a works rider taking the British Experts Trial in 1937.

Enfields were machines to buy and ride, to use in the occasional weekend sporting event and for this they were good. No complex overhead camshafts for them, nor the all or nothing gamble of the race track. The firm was run by men who knew only too well the conservative nature of the market place and allowed others to build the glamour machines, receive the rapturous reports in the press, and then the empty order books. Motorcycle buyers asked for, inspected and argued the merits of multis, ohc, blowers, and all

Below **The experimental 350 cc side valve twin with enormous case enclosing both chains. Only this prototype was built**

Above **Loading a batch of model CO machines on to an RAF lorry**

the rest but went out and bought simple singles, or the Triumph twin which managed to look like a single. The after-effects of the depression were then still too close to allow many to indulge in glamour buying; it was transport that was needed—straight-forward and easy to repair.

Enfield provided what was wanted and what the public would put its money down for. The machines did gain a reputation for being 'built like a gun' and, although in some quarters this was viewed as indicating excess weight, it became a friendly joke. Perhaps the firm should have showed a rifle in their trademark and not a cannon. They were a make that, like BSA, was looked down on by owners of machines with a racing pedigree but, again like BSA, sold well into another and much larger market. Never exciting but many non-sporting riders would choose the Enfield for just that reason—it could be trusted to get them to work without a plug change *en-route*.

For all that, Enfield did dabble with the unusual over the years and felt the water with machines such as an in-line four built as a one off in 1920, a triple two stroke, sleeve valve singles, and a transverse 500 cc vee twin with shaft drive built in 1934 but too expensive to put into production.

With all this wealth of experience Enfield was

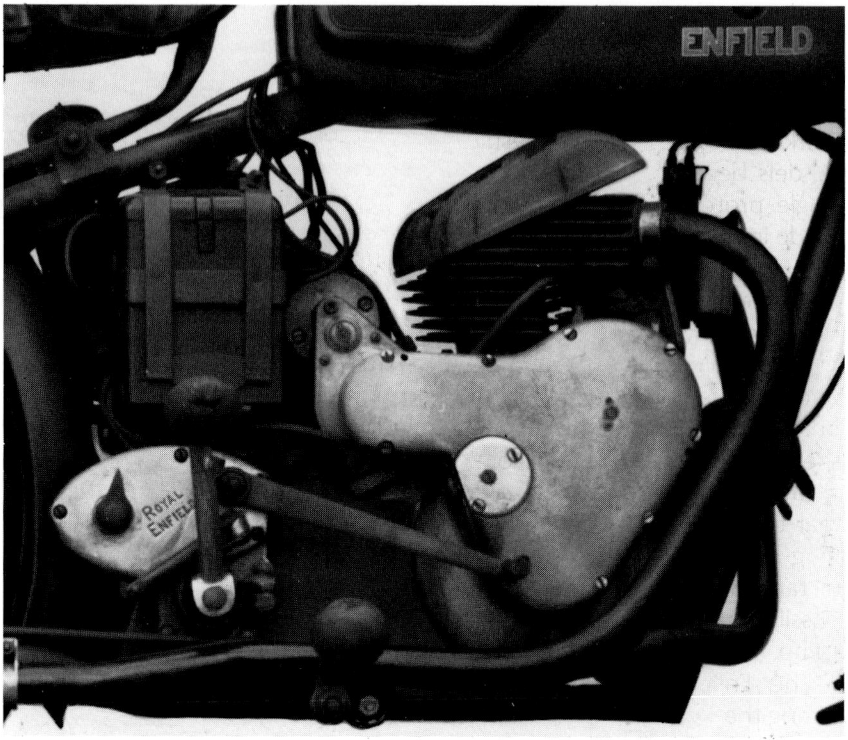

Engine of the WD twin. Inclined cylinders, horizontal fins, front carburettor and combined seat and rear mudguard unit, easily detachable

called on for machines during the second world war and responded with a baby two stroke and two versions of their 1938–39 350 cc models, one with side valves and the other overhead. In all, over 55,000 machines were to be supplied.

The two four strokes were very similar to one another using the same cycle parts and bottom halves to the engines. Only the head and barrels with the accompanying detail fittings varied, along with the gearing which was lower on the sv machine. In most cases an Albion gearbox was fitted but during the war, on contract 294/C/19870, the machines were fitted with a Burman unit which looked much the same but used its own positive stop mechanism. The cush drive clutch centre normally found in the Albion box was replaced by a solid centre and machines so fitted carried the identification WD/CO/B, the additional letter 'B' standing for Burman of course. It is believed that only the overhead valve CO model used the alternative box and all side valve Cs had the Albion fitted.

Both types of machine had the Lucas mag-dyno with regulator beneath the saddle, Amal carburettor with gauze air cleaner, and the accoutrements of military service such as canvas pannier bags on frames, rear carrier, headlamp mask and khaki paint. The simple air cleaner was something of an economy measure for pre-war Enfields had been fitted with a large box-like cleaner mounted where the rest of the industry put their oil tanks—on the seat tube.

The baby two stroke was the result of an earlier approach from the continent and, although barely developed in 1939, it came to be adopted by the airborne troops later in the war. It was to become the 125 model, described in a later chapter.

Enfield also built other models for use by the services and among them were a 500 cc sv single, used by the navy, a prototype lightweight 350 cc, also with side valves, and a 250 of which about 1000 were built and used for training purposes. During the war some experimental work was carried out and resulted in a few WD models being built with telescopic forks and a single prototype 350 cc sv twin. This last was made in response to a ministry specification and had a parallel twin engine with the valves, carburettor and exhaust pipes at the front. The cylinders were laid back a little but the fins remained parallel to the ground which gave it an odd appearance. The gearbox was an Albion and both chains were enclosed in a vast single case, the rear chain being tensioned by a jockey sprocket. The rigid frame had girders and a fat rear tyre while the electrics were reduced to a mag-dyno, battery and simple lights. A separate oil tank and dry sump lubrication were adopted, possibly to lessen the consequences of a cracked sump.

The Enfield machines did sterling service during the war and, along with the other makes, were sold off to the trade in large batches just after it. Thus the pages in the backs of the two weekly magazines carried block adverts with the line drawings of the models, and a machine-hungry public snapped them up. At that time new machines were mainly destined for export and took an intolerable time to reach the home market.

Enfield themselves joined in this exercise and produced factory rebuilt ex-WD models complete with girder forks. Both side and overhead valve 350s were produced and finished in black with gold lining, the wheel centres also being in black. Production was sufficient to warrant their inclusion in the sales catalogue along with the new civilian machines.

The home market was booming for thousands of men had been trained to a knowledge of mechanics during their service, had experienced the usefulness of motor transport so craved their own vehicle, and all had a gratuity burning a hole in their pockets. The problem was to find a machine to buy, so prices of pre-war models zoomed, restrictions and covenants surrounded the supply of new ones, and the released surplus of army machines filled a gap. In time new models did reach the home buyer and the crude military machines were driven into the ground, although that was never easy with anything capable of withstanding service use. Or they were just left to rot away until genuine ex-WD machines became rare and then valuable in the resurgence of interest in older machines that occurred in the late 1970s.

The all-alloy 1939 Bullet engine. Unlike postwar units the oil filler is in front of the cylinder and the rocker covers are ribbed

Before that was to happen Royal Enfield were to produce a variety of models and go through many phases.

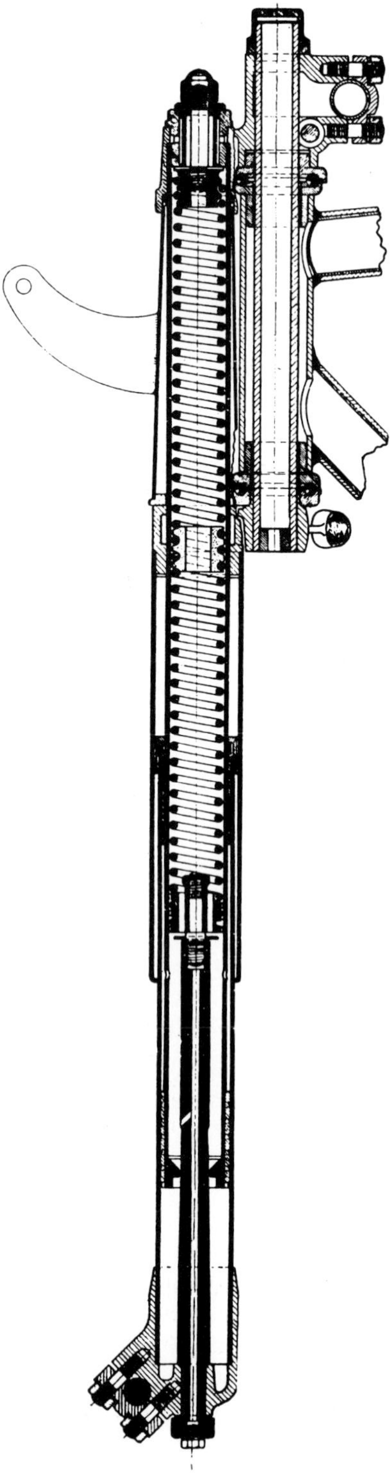

2 Big Single

When the war ended the most urgent need was production. The production of machines to satisfy the cries of the politicians to export and production to meet the demands of a home market desperate for its own transport. There was no time to ponder on the whims of design, to consider novel ideas or indulge in thoughts of new production methods, machines in quantity were what were needed despite all the problems of material supply and obstructive restrictions imposed by the wartime bureaucracy and taken over with keenness by its peacetime counterpart.

In spite of the problems and the regulations, in spite of the government officials, in spite of the turmoil of change, machines were built. Inevitably they were little changed from the wartime ones and in most firms were one or two pre-war models with the welcome addition of telescopic front forks in some, but not all, cases. As far as finish was concerned yesterday's khaki was discarded and general purpose black poured into the spray gun. That black coped for the years while the industry struggled back onto its feet.

At Enfields they announced a range of three models in November 1945, one based on the wartime 125 cc two-stroke and covered in a later

The telescopic front forks fitted to the models G and J when announced in 1945. At the top is a breather with filter

chapter, and two four-stroke singles. Aside from obvious capacity alterations and one or two minor points these were identical and based largely on the pre-war and wartime ohv machines.

The models were the G with 70 × 90 mm bore and stroke and 346 cc capacity, and the J which had an 84 mm bore while retaining the 90 mm stroke to give 499 cc capacity. Compression ratios were 5·75 and 5·5 : 1 respectively thanks to the poor quality Pool petrol that was all that was available in those days, although alternative pistons with higher ratios were promised as soon as better fuel was available.

The alloy piston was attached to an alloy connecting rod by a fully floating gudgeon pin retained by circlips and which worked directly in the rod small end eye. The big end bearing was an Enfield feature with its origins back in the middle 1930s. In that period Enfield had experimented in the TT with an alloy rod with pressed in and ground hardened steel sleeve running on a plain floating aluminium alloy bush which in turn worked on the steel crankpin. This bearing worked but only if it had a really copious supply of oil and plenty of running clearance. The second requirement gave a very noisy engine and also made it hard to hold much oil pressure across the bearing so the bush was changed for one in steel that was faced with white metal on both sides. This enabled it to run with much smaller clearances and was adopted by a

A 1941 service single fitted with prototype telescopic front forks and a well valanced front mudguard

number of Enfield engines from then on, including all the wartime 350s. This then was the big end bearing used on the postwar 500 cc model with a pair of washers to control the side float. On the 350 a return was made to the use of a special alloy bush that did not require to be coated with white metal.

The crankpin was formed with parallel ends which pressed into the flywheels and were pulled up by nuts. The mainshafts were taper fitted into the flywheels, keyed and also secured with nuts. The complete flywheel assembly ran on a double row caged roller race on the driveside and a single caged roller bearing plus a bush on the timing side. The crankcase was split on the

vertical centre line of the cylinder which was attached to it by five short studs, the head fixings being separate and four in number. The case halves fitted together around the flywheels and the inner walls that enclosed the crankshaft were themselves encased by those of the outer case which carried the lubricating oil. This gave a substantial appearance to the crankcase for it extended forward of the cylinder to a noticeable amount. This extension carried a screwed in filler neck on the timing side which was closed by a quarter turn cap complete with level dipstick. Two filter plugs were screwed up into the underside of the timing side crankcase.

The extension of crankcase behind the flywheels was put to good use by mounting the mag-dyno on it. The drive to it was contained in a timing chest of which the inner part was formed in one with the right crankcase half. The drive itself was by a series of gears running on fixed spindles pressed into the wall of the crankcase and supported at their outer ends by the timing cover which enclosed the drive. Six gears in all were used with the crankshaft pinion driving the exhaust camshaft gear which in turn drove the

Above **A factory rebuilt model CO complete with girder forks. These were sold alongside the newer G and J models**

Below **The famous Enfield twin oil pumps and filter. Their shape and layout typified the marque's timing cover for many years**

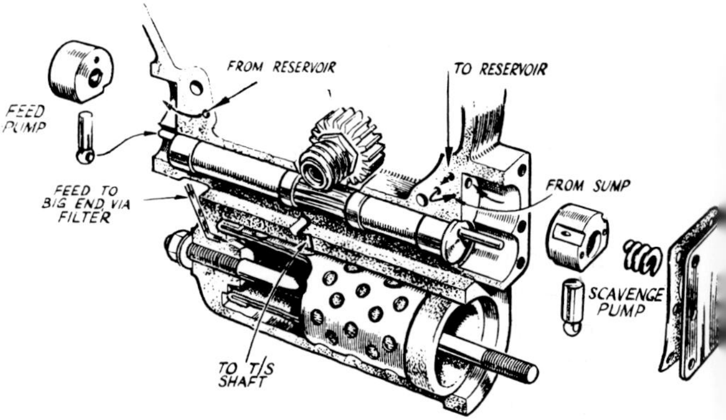

The 1945 model G which had few changes in its life. No pillion seat was fitted and the speedometer was always an extra

inlet and this in turn drove the magneto via two idlers.

The valve gear was quite straight-forward with flat faced tappets working in guides pressed into the crankcase. Above the tappets rose a pair of pushrods through a tunnel cast in the cylinder. This had an aperture giving access to the screw adjustors at the lower ends of the push rods and this was sealed by an alloy plate secured with a single arm wingnut. The push rods passed on into a tunnel in the cylinder head and the head to barrel joint was sealed round them by extra washers. The main joint was by spigoting the barrel up into the head with a copper jointing washer but, to ensure that oil did not leak from around the push rods, special measures were taken. For each rod a ferrule was screwed into the cylinder head so it hung down into the barrel tunnel to ensure that oil drained down into the tunnel rather than out through the joint. To ensure that nothing worked itself out via the ferrule threads, the barrel top was counterbored to take a soft washer and this sealed against the outside and the thread of the ferrule.

The top of the cylinder head was machined off flat and fitted with a number of studs. On these dropped the rocker housings which comprised a single lower split housing and two separate top halves. The rockers were forged in one and, with everything assembled, were enclosed by an alloy top cap secured by a single nut. The valves ran in pressed in guides, were fitted with hardened end caps and restrained by duplex springs retained by a collar and a pair of cotters.

The lubrication system was dry sump and based on two plunger pumps driven by a shaft that ran fore and aft in the timing cover. This was driven by a worm gear which screwed to the end of the crankshaft and also retained the timing pinion on its keyed and tapered location. The pumps were located at either end of the shaft under small covers held in place by six small screws. Under each cover were two counter-bored recesses which overlapped to give a figure eight format. The pump shaft emerged from the lower one with an offset drive pin on which hung

ROYAL
ENFIELD

AIR FILTER
2,000 MILES
ROYAL
ENFIELD
R.E./1978

Left **the 1945 model J, very hard to distinguish from the smaller G as all cycle parts were identical**

Left below **1946 model G cutaway for training and exhibition purposes to show the inner workings**

Right **Period Enfield advertisement run in 1946 with various headings all emphasising the return home of the servicemen**

the pump plunger. This worked in a body, which was basically circular with a single flat, and this oscillated in the upper recess and was forced against its inner wall by a compression spring. The two pumps were formed in the same manner with the rear one the delivery side and the front one the scavenge. Below the pump shaft a further section of the timing cover formed a housing for the main feed filter which was inserted into place from the front and sealed by a cap secured with a one-armed wingnut.

This layout of timing cover, oil pump housing and filter was to characterise the Enfield single for many years as it had done in pre-war days.

The upper recesses in the pump housings contained drillways which formed part of the oil circuit along with suitable holes in the timing cover and crankcase. The operation of the pumps is complicated because both are double acting and thus pump in both directions. The feed pump is ported so that the pump piston and cylinder supply the big end while the back of the piston is used to pump oil to the rear of the cylinder wall. In both cases the oil is supplied from the reservoir via one of the small filters in the underside of the crankcase. The feed to the cylinder was straight through drillways to the rear thrust face of the piston but that for the big end was forced through the main felt filter under the pump and into a supply quill sealed into the mainshaft and so to the big end journal.

On the return side the pump collected oil from the sump via the second small filter in the underside of the crankcase and both sides fed into the same return line. This line ran back to the sump but contained a simple ball valve to generate enough pressure to force a little of the returning oil along an alternative drillway to a union in the top face of the crankcase just behind the filler cap. This was connected by an external pipe to the cylinder head and hence to the rocker spindles. From there the oil drained down the push rod tunnel and through grooves in the tappet guides into the timing case. It was then picked up by the timing gears to lubricate them and, on being carried round by the gear teeth, was eventually forced through a passage high in the timing case into the rear part of the oil tank.

The tank itself was vented at the cap and the

rear part vented to the front via a groove machined in the joint face. For 1947 a vent was arranged immediately below the mag-dyno which was sealed to the back of the timing case with an assembly of felt, cup and corrugated washers and had its drive gear fixed to a tapered shaft.

An Amal carburettor was fitted and supplied with air through a large cleaner element housed in a box mounted on the rear down tube. The exhaust pipe swept down on the right side of the engine to a low-level silencer bolted to the rear chain stays. Within a few months an upswept system, still fitted on the right, was offered as an option and came complete with a heat shield clipped to it.

On the drive side a single strand chain in a pressed steel chaincase transmitted the power from engine to clutch. The chaincase was quickly detachable and equipped with oil drain and level

A model G from the 1946/47 period. Slightly brighter finish, very few detail changes but still no rubbers on foot controls

screws as well as a combined filler and access hole which allowed chain tension to be checked. This hole was blanked off by a cap held by a sprung cross arm.

The clutch was of the four-plate type with cork inserts clamped by three pressure springs. It drove a separate four-speed Albion gearbox with positive stop foot-change pedal on the right. This box pivoted in the engine plates to provide primary chain adjustment and was of conventional English layout with the output sprocket carried on a sleeve gear running inboard of and concentric with the clutch.

Internally the box differed from the usual layout in that the moving gears were all moved as one by a single fork. To enable this to occur the two centre gears on the mainshaft were cut on one piece of steel and splined to the shaft, while the corresponding layshaft gears were separate but constrained to slide together

The J2 model with twin port cylinder head and in this case upswept pipes and silencers, also narrow mudguards

ROYAL
ENFIELD

although rotating at different speeds. These engaged a series of dogs on the layshaft to provide the desired gearchange sequence, the only fault with the arrangement being the rather high inertia of the parts to be moved at each gearchange. This could lead to the box over-changing if hurried too much.

The positive stop mechanism had one unusual feature in the form of a neutral finding lever positioned above the main gear lever. This could be used to select neutral from any gear except first by merely pressing it down to an adjustable eccentric mounted stop. It was to be an Enfield feature for many years.

The gearbox also carried the kickstarter mechanism on the right and this turned the layshaft bottom gear in the usual English style. The final drive was on the left by chain and this was protected by a guard over the top run.

The engine and gearbox were held together in plates and mounted in a massive cradle frame with single top and down tubes, the latter splitting into a duplex cradle under the engine and running right back to the rigid rear wheel. Duplex upper chain stays and a single saddle tube completed the structure which carried the new Royal Enfield telescopic front forks in the substantial head bearings.

The forks used long internal springs secured to the forks at both ends so that they could work in tension on rebound as well as in compression. The lower legs worked on the outsides of the stanchions and contained a two-way hydraulic damping system that was taper controlled to be progressive in action. At the very top of each fork leg was a small filter to ensure that any air taken in during the movement of the fork was kept clean. The tapered damper control rod was anchored to the lower fork end and hollow. Its bottom end was closed with a set screw and it

A Paris policeman on his 1949 model G comparing notes with a civilian owner. No chrome plating on tank and continental style pillion

was cross-drilled to establish the oil level in the forks correctly. The owner simply poured oil in and let it find its own level before replacing the lower screw.

The forks were made in both solo and sidecar forms in due course by changing the yokes to alter the trail and by fitting heavy-duty springs in the sidecar type. The solo yokes had a front bridge that was nearly straight while the sidecar ones were vee-shaped to increase the fork offset. In either case the fork ends were offset to place the wheel spindle in front of the fork tube centre which reduced the inertia effect of the parts. The spindle was held in place by caps fixed to each fork leg.

The wheel spindles were both solid and the hubs ran on a pair of ball races. The rear hub also contained the Enfield cush-drive which was a transmission shock absorber of the three-vane type which used six rubber blocks to provide the resilient member. Simple single leading shoe drum brakes were fitted to both wheels and a ribbed alloy back plate carried the front brake shoes. At the rear a knock-out spindle was soon adopted to allow removal of the rear wheel without disturbing brake or rear chain. The rear brake was rod operated directly from the substantial pedal mounted on the left.

Both wheels had 6 in. brakes and 19 in. rims with a 3·25 in. front tyre on both models, the same size for the rear wheel on the G but a 3·50 in. for the J. Mudguards were large and sensible. The front one was sprung, very deeply valanced, and without stays, being supported by a pressing fixed to the lower fork yoke. This

Above **The model G for 1950 in a publicity picture taken in September 1949. Note that the small toolbox is no longer fitted**

Right **A model G pictured in mid-1950 in a special finish and with the flying wing style of tank motif**

pressing was attached to the top of the mudguard and also served to carry the then mandatory front number plate.

At the rear the mudguard was in two sections, one fixed and covering the sector between the chainstays and the remainder, together with supporting stays and rear number plate, readily detachable by slackening four nuts and disconnecting the lead to the rear lamp. This made wheel removal exceptionally easy and, in fact, punctures could be repaired without taking the wheel out. The detachable section was supported by a rear stay that ran straight back on each side and a pair of lifting loop stays. The loop stays ran up to the guard, followed its contour for a distance and then bent back in to the wheel centre again. In each loop was fitted a small toolbox and a third, larger box went in between the chainstays on the right side behind the gearbox.

The machine was finished in black with a chrome-plated petrol tank with frosted silver panels and the maker's name in bright red. Behind it went a saddle and the rear guard could carry a pillion pad, pillion rest lugs being built into the frame. The front forks carried the headlamp and the top yoke the speedometer which was driven from the rear wheel. The handlebars were described at the time as being unusually narrow. Two stands were provided, one of which pivoted on bolts in the rear fork stays and held that wheel clear of the ground. The other was a prop stand which bolted to a lug on the left lower chainstay and at rest was held in a clip to point rearwards. It was really mounted too far back for the best but worked well thanks to its length, sharp end and guard which held it on muddy ground.

For electrics, Enfields used standard Lucas parts with the regulator beneath the seat, the battery on the left to balance the air cleaner and control switch with ammeter in a small mounting panel in the back of the headlamp. The horn was bolted to the front upper engine mounting on the left and the headlight was the rather insipid

device of its time with flat glass and bulb that could be moved for a better or different focus.

The machines went onto the home market at £95 for the 350, and an extra tenner for the 500, both prices plus purchase tax. To this had to be added the optional, but mandatory in law, extra of the speedometer at £3 10*s* 0*d* plus tax.

By 1947 an extra large capacity petrol tank was available and the exhaust pipe had a finned collar clamped to it at the cylinder head. That year *The Motor Cycle* road tested a model G with big tank and found it a straightforward tourer with some excellent rider points which were fully appreciated. They made the performance as 69 mph in top with speeds of 56, 39 and 26 available in the gears and found that it cruised at a mile-a-minute without any tendency to overheat. At that time this was a respectable speed and one that gave good averages for most journeys. The machine handled well, the brakes worked nicely and, even if not over exciting, the model provided good reliable transport which was just what was wanted in those times.

For 1948 the home range was curtailed to the model G with the larger machine reserved for export only. As production was the prime essential in those days changes were few and only concerned a slight alteration in the gearing, the provision of pillion rests as standard, and improvements in the finish. This last affected the lower fork legs which were chrome-plated while the fork ends were polished. The wheel rims were also brightened with chrome plating with the centres in black.

There were few changes for 1949 either although the model J became the J2 with twin port head and an exhaust pipe and silencer on each side of the machine. It was still export only and it, along with the 350, lost the ribs on the front brake backplate to make cleaning easier. It was hardly surprising that few changes occurred as the works was well occupied in introducing two more exciting models, described in later chapters.

In 1950 the J2 came back onto the home market in solo or sidecar specification, the latter with the alternative fork yokes, heavier springs and lowered gearing. Both singles received longer dynamos with greater output which in turn allowed the use of headlamp bulbs with more power. The rear mudguard stays were cleaned up by substituting a single central lifting loop in place of the two fitted before at the sides. Also only one toolbox was fitted on the model G on the right between the chainstays while the J was provided with one on each side.

The forks were changed a little in 1951 with the ends becoming forgings which were welded to the fork legs. At the same time an unsprung front mudguard supported by three stays on each side was adopted. The J2 also had the top end of the forks changed when set up for sidecar use. A new top yoke was cast in alloy to carry the speedometer as well and with a new forged bottom yoke gave the required trail. A steering damper was also provided with the new arrangement and the machine in sidecar specification cost a few pounds extra.

There was to be little further alteration to the two simple rugged singles which continued with very few changes into the mid-1950s. In 1952 the G model was fitted with the cast alloy top fork yoke which also carried the speedometer and both models had the prop stand altered to incorporate an over-centre return spring.

The finish had been changed the previous year due to the world wide shortage of nickel which forced a reduction in the area of chrome plating. On the Enfields this was overcome to some extent by painting the petrol tanks black with gold lining and reducing the plating to a chrome trim strip running along the tank top.

The lining on the tank went by 1953 when an embossed motif was used on the tanks of the whole range and the two rigid singles continued in black. They had a small change to the brakes to allow the cam spindles to float but that aside continued as before.

In 1954 the model G was dropped, effectively superseded by the 250 cc Clipper model, but the J2 continued to cater for those riders who still wanted a rigid frame, mainly for sidecar work. The model ran on into 1955 but the market for that type of machine was contracting so Enfield decided to concentrate on their other models and the solid single was laid to rest.

A 1954 model J2 engine and gearbox unit fitted into a swinging fork frame. Not listed as standard

3 Bullet

While the solid singles filled a need just after the war not even the Enfield adverts could claim that they were exciting machines and, as far back as 1933, they had overcome this deficiency by introducing the Bullet range. These were more sporting models, often loosely based on the tourers but sometimes going out on their own more adventurous paths.

After the war, and as soon as the immediate calls for transport had been met, such models were needed to cheer riders up so Enfield obliged with a new 350 cc Bullet. The first prototypes of the new model were seen in the 1948 Colmore Cup trial ridden by the works team trio and caused a sensation for they were fitted with swinging fork rear suspension, then unheard of for trials use. While the trials debut only had a limited success the machines soon showed their form and won golds late that year in the ISDT and were part of the winning British Trophy team.

The prototype machines had many points in common with the existing singles but these related more to detail parts such as the oil pump rather than the overall layout. The engine still had a dry sump oil tank in the crankcase and drove an Albion gearbox but it was all arranged in a much more compact form to suit competition use.

The 1948 Bullet engine cutaway to show the train of timing gears and sleeve in the original alloy barrel

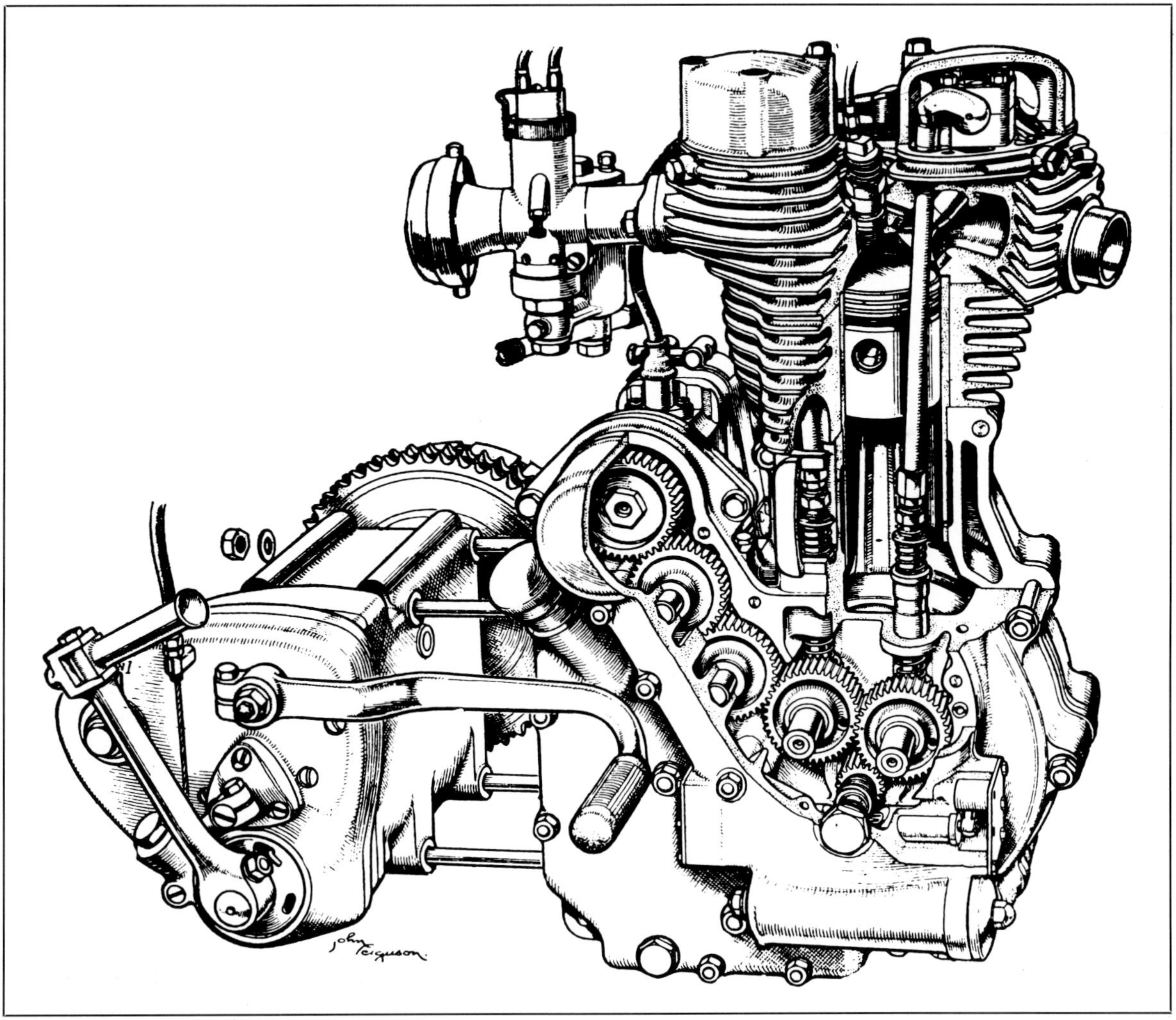
John Ferguson.

The Bullet prototype in March 1948. No oil tank and raised saddle give an odd appearance

Bore and stroke remained at 70 × 90 mm and the big end bearing was that from the model G with its alloy bush, steel pin and big end eye, and alloy connecting rod. The layout of the timing case with its row of gears running up to the magneto and the cover containing the duplex oil pumps and the filter were also the same. Completely different was the crankcase for the oil was all carried behind the flywheel compartment and the filler cap neck rose from this chamber behind the timing case by the magneto. This move greatly shortened the engine as the front of the crankcase became similar to any other single, although it did retain the oil feed supply to the rockers.

The crankcase was split on the vertical centre line of the cylinder and this item was in alloy with a cast-in liner. Unlike the touring engines it was very deeply spigoted into the crankcase, so far in fact that the barrel joint was above the tappet adjustment inspection plate and that item became attached to the crankcase. Otherwise the timing gear was the same with flat footed tappets working in pressed-in guides to move alloy push rods with hardened ends and screw adjusters. Detail alterations were made to the

Bullet prototype with Charlie Rogers (left) and Jack Booker when the machine was first seen in 1948

The scrambles version of the 1949 Bullet with open pipe but hefty standard front mudguard support

timing pinion and oil pump worm, both of which were machined as part of the mainshaft which was keyed into the flywheel and held to its taper by a nut. The top of the timing cover was also changed slightly being bulged out to allow an auto-advance unit to be fitted to the magneto drive pinion if desired. Although a racing magneto was fitted to the platform behind the cylinder the available space was sufficient to enable a mag-dyno to be fitted when desired.

Inside the crankcase the flywheels were polished steel and turned in a plain phosphor bronze bush on the timing side and two ball bearings on the drive. The drive side crankcase was webbed internally to ensure that the bearing housing was rigid and well supported.

The head and barrel were held down by five long studs in the crankcase supplemented by a case to barrel fixing between the pushrods and another above it from barrel to head. Like the barrel, the head was in alloy and had cast-in iron valve seats. The valves had hardened caps and worked in pressed-in phosphor bronze guides and were restrained by duplex valve springs retained by collars and cotters. The top of the cylinder head was machined flat and contained two separate wells for the valve gear. On each sat a split cast iron bearing in which the one-piece rockers pivoted, each valve assembly being enclosed by a polished alloy cap.

Between the caps on the left was the sparking plug and on the right a decompressor valve which communicated the released gases into the exhaust port. The head was recessed for the barrel to spigot into it and had provision for a flange mounted carburettor. The exhaust port carried a stub for the exhaust pipe and, in place of the single oil pipe connection of the G, the supply line ran up the front right corner of the barrel to the head joint where it turned back to the centre of the push rods before rising again and then splitting at a tee-junction. Each arm of the tee ran round the head to a union which supplied the rocker spindle on that side.

Lubrication was standard Enfield and the breather worked through the drive side mainshaft, which contained a non-return disc valve, but not into the chaincase. The engine sprocket nut was extended into a recess in the chaincase cover and from this a drilled passage led down through the cover wall to the open air.

It would have been easy to have lost the

Left **The production road model Bullet for 1949 fitted out with full equipment**

Right **1950 Bullets on duty with the Leicester police. Legshields added for weather protection but otherwise standard**

compactness gained from the new crankcase design by fitting the gearbox in conventional plates, but Enfields did not do this. A flat face and four mounting studs were provided on the back of the oil compartment of the crankcase and the gearbox was bolted directly to that. This gave a semi-unit construction with fixed chain centres so the primary drive was by a duplex chain with slipper tensioner running in a two-piece aluminium chaincase. This was sealed at the joint by a round rubber strip which fitted into a groove in the rear section, while the outer case was retained by a single fastening.

The clutch had Ferodo discs riveted to the sprocket drum and three plates with cork inserts. It was operated by a quick thread worm rather than the usual Enfield long lever, the assembly being attached to the outside of the gearbox outer cover. The box was standard Albion with the gear pedal pivot high up, the kickstarter working on the layshaft and a gear indicator pointer on the cover. No neutral finder lever was fitted and final drive was by chain on the left.

The engine and gearbox unit was mounted in a new frame with swinging fork rear suspension. It had single top and down tubes, the latter terminating in a small cross tube to which the tops of the front engine plates were bolted. These were combined with a stout undershield which ran back under the engine to further plates and a cross bolt which went through the bottom of the gearbox shell. A single saddle tube supported a further cross tube which had end plates to attach to the top of the box.

The rear frame loops ran from the junction of top and saddle tubes out and back to support the tops of the suspension units before turning down to pass behind the fork pivot point and on to join the rear engine plates. Bracing tubes ran from just above the fork pivot to the saddle tube

and plates in the resulting corner carried the fork itself. This was based on a malleable casting into which were brazed the two tapered fork legs and the assembly pivoted on plain bronze bushes working on a steel spindle.

Wheel movement was controlled by hydraulically dampened spring units with covers and the wheel carrier plates at the fork ends supported these on Silentbloc bushes. The left plate was extended down and slotted to provide a brake anchorage point.

At the front of the machine were standard Enfield telescopic forks and both wheels had 6-in. brakes. The rear hub contained the standard cush-drive and was moved for chain adjustment by serrated snail cams fitted on each side of the wheel spindle and which worked against fixed pins.

On the prototype, tyres and gearing were selected for trials use and there were a number of points that were still under development. Despite this the general reaction was most favourable and sure enough, when the 1949 range was announced, it contained a Bullet complete to road going specification. This was soon followed by details of a whole range of Bullets which were available in trials or scrambles trim as well as the road one. This range was achieved by using the same basic parts but varying the detail specification of compression ratio, exhaust system and tyres along with other points and the general equipment fitted.

Relatively little had to be changed from one machine to another and all were very much as the prototype first seen early in the year. One item that did change was the cylinder which became cast iron. The weight saving of the alloy job with liner was minimal with such a deep spigot so the complication was unwarranted. Cooling was augmented by continuing the barrel

Line drawing of the 350 cc road Bullet in 1951. Used by many riders in this and its competition forms, all with the oil tank in the crankcase and semi-unit construction

ROYAL ENFIELD
MOTOR CYCLING
COPYRIGHT

Foretaste of the future—Mr Sankaran of Madras Motors checks out a 350 cc Bullet built for the Indian Army in 1951

fins onto the crankcase mouth. The road model used the front mudguard, headlamp and air cleaner from the model G and was fitted with a very stylish exhaust system with uptilted silencer. It was the only one of the three to have the neutral selector fitted and came with a saddle and pillion rests, the pillion pad being an extra which fitted to the rear mudguard. This guard was supported by two long stays which ran back from the frame and were welded to curved stays that ran round the guard to the subframe. As on the rigid models, the major part of the guard could be removed by slackening a few nuts.

The road model was equipped with two toolboxes, one each side, and two stands, one centre and one prop. The whole appearance was compact and sporting with the tank in chrome plating with frosted silver panels and the company name in red. A neat detail touch was the adoption of plated domes to the top of each fork leg.

The trials model had a Lucas racing magneto, no lights and a bulb horn, although it could be ordered with the roadsters mag-dyno and lights if required. A wide ratio gearbox with folding kickstart lever was fitted and a high level exhaust system ran back past the tappet cover to the normal reverse cone-style silencer. The carburettor had an inlet shield in place of the air cleaner and other alterations affected the tyres, mudguards and use of security bolts in both wheels.

For scrambling, a straight through exhaust pipe was fitted, even lighter mudguards, appropriate tyres, a close ratio gearbox and one from a choice of three pistons to give ratios of 7, 8·5 or 10·5 : 1. All three models were priced at £135 0*s* 0*d* plus £36 9*s* 0*d* tax.

Both *Motor Cycling* and *The Motor Cycle* had trials models to test during the spring of 1949 and reported well on the ability of the machine to deal with typical trials sections of the day. Both magazines found the rear suspension a decided asset and felt that Enfields had started a trend that just had to continue—which it did.

There were few changes for 1950 but the silencer was altered to a cylindrical shape and the dynamo to the longer type with greater output. The front forks received a cast alloy top fascia that carried the speedometer and the front mudguard became shallower with reduced valances, although it remained sprung.

Motor Cycling road tested a Bullet in February and reported favourably on it in all aspects but were especially complimentary about the suspension. This was certainly understandable for at that time most machines had either rigid or plunger frames so riders had to rely on their saddle springs for comfort. The test Bullet proved capable of 74 mph with speeds of 66 and 46 being available in the higher intermediates. Thus it could cruise at 60–65 mph all day and, under such conditions, was able to return over 80 mpg. The brakes worked, the machine handled and maintenance was very easy to carry out, thanks to some well thought out features. It was not surprising that they sold well and owners liked what they had bought.

The only change for the road model in 1951 was the adoption of the welded-on, forged steel, fork ends and the unsprung front mudguard with stays that was common to the range. The trials model also had its petrol tank slimmed down to 2 gallons, its compression ratio dropped a touch to 6·0, the gearing raised a little and the exhaust pipe run modified. The pipe was curved on the lines of the works machines so that it ran over the top of the timing cover so allowing that item to be more easily removed. It still obscured access to the tappets, but improved it for the oil filter. The pipe diameter was also reduced as was the size of the carburettor from 1 in. to $\frac{15}{16}$.

About this time the works type alloy barrels began to become available so some of the competition models appeared with these, although the road machines continued with the cast iron type. This became standard in 1952 and the trials model also had its footrests moved back a little and the right made to fold up to allow the kickstarter past it. The works used a special curved lever which curled under the footrest to avoid this problem and were well copied. On the road model the breather outlet on the chaincase gained a banjo union, the prop stand an over-centre spring and dualseat was made available as an option on new machines.

For 1953 there was a new Bullet of 500 cc and a number of internal improvements to the engine of the 350. The larger Bullet looked just like the smaller one and used nearly all the same cycle parts. Alterations concerned the swinging fork which was strengthened, a larger rear brake and the fitting of sidecar lugs. As with the model J2,

The trials model 350 Bullet. This is a 1952 version although it has the 1951 style of petrol tank panelling

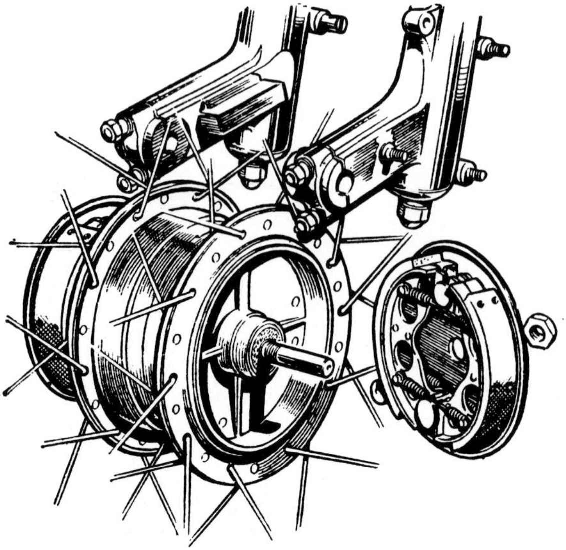

Above **The 500 cc Bullet, also called the model JS. This is the 1953 version and it has a fully painted petrol tank**

Left **The extra long end lugs on the forks gave the sidecar trail and the dual brakes more stopping power. Not always used together**

Right **A 1954 trials Bullet of 500 cc. Centre stand, side spike and rear brake pedal certainly 'built like a gun'**

Left **The 500 cc scrambles Bullet as prepared for the USA in 1954 with high bars, silencer, lights and horn. Really trail or enduro specification**

The 1955 Bullet as sold for road racing with rearsets and GP carburettor. The bars, seat and filler cap are less than suitable

the larger Bullet was available with forks and gearing to sidecar specification together with a steering damper. Both sizes of machine were fitted with floating cam spindles on the front brakes and a new Lucas stop and tail lamp with which came a round reflector mounted at the base of the rear number plate. Less popular was the adoption of the Lucas headlamp with underslung pilot lamp, a device that proved singularly useless and which riders only used when parked.

The 500 Bullet engine was based on 84 × 90 mm dimensions and, while it followed the lines of the 350 in many areas, it did incorporate a number of changes which went into the 350 as well. Compression ratio was 6·2:1 and a split skirt piston was fitted to further reduce engine noise. The crankshaft had the same bush big end bearing fitted but ran on a new arrangement of main bearings. On the timing side a roller race was fitted inboard of the phosphor bronze bush, while the drive side was supported by two roller bearings in the 500 and a roller and a ball bearing in the 350.

The detail of the timing mainshaft was as in the G model with separate timing pinion secured on its key and taper by the oil pump worm which was internally threaded. On the drive side the mainshaft became solid and machined with splines to carry the engine sprocket as the engine breathing through it had tended to lead to sticking clutch plates when the oil fumes leaked into the chaincase. With the mainshaft blocked engine breathing was looked after by a clack valve assembly bolted to the left crankcase just below the cylinder.

The lubrication system was changed, although still based on the same idea with two double acting plunger pumps, one at each end of a fore and aft shaft, with oil filter chamber below the pump housing in the timing cover. One noticeable change was that the end covers of the pumps became oval in shape and were only retained by four screws. The two small filters in the underside of the crankcase were retained.

In July *The Motor Cycle* road tested the 500 and found it a good 70 mph cruiser with the torque and power curves that allowed it to roll on uphill and down dale with little speed variation. The maximum came out at 78 mph with 70 and 53 available in the higher gears. There were a few snags for the exhaust pipe blued rather a lot in 1500 miles, albeit hard ones, and oil tended to leak from the filler cap. It was noted that the outer case had to come off to check the primary chain tension, but with the comment that this was easy enough as only one nut held it. The need to drain and refill the case with oil was not mentioned. The riding position was good but would have been better if the gearbox had allowed the footrest to come back further and if it had been possible to change gear by pivoting the foot on the rest. The gearchange itself was good, only the high pivot point of the lever spoiling the effect. The clutch was a little heavy and tended to slip under prolonged hard riding. The machine steered and handled well, while the brakes proved to be up to their work. The centre stand was easy to operate but did not raise either wheel from the ground. All told it was a machine for use day in and day out without temperament.

1954 brought a styling change and a rear suspension improvement. The major alteration to the appearance was to the area at the top of the forks and achieved with a die-cast alloy cowl Enfields called a 'casquette'. This enclosed the tops of the forks and extended back to the steering head. It carried the headlamp unit and above this and on each side a small pilot light as well. The top surface acted as a mounting for the speedometer, ammeter, and light switch, while also containing grommets to guide clutch and brake cables out of sight. The machine appearance was also altered by mounting the silencer so that it was parallel to the ground, although some owners felt that it lost something without the cheeky uptilt of previous years. At the rear end suspension units were changed to proprietary ones and this increased the rear wheel movement.

In 1955 the competition Bullets were listed as both 350 and 500 cc models and in both cases with a short circuit racer as well as the more usual trials and scrambles machines. The racer was fitted with a straight-through exhaust pipe, the standard dualseat, rather high bars and, for footrests, utilised the normal pillion rests. Thus it lacked the purposeful air of such contemporary models as the Manx, 7R or Gold Star. It did have a GP carburettor but used the same petrol tank as the trials model. It was not to be taken up in any numbers, as the BSA Gold Star was much more successful for either Clubman or short circuit racing if you could not afford a Norton or AJS.

The road models had some changes made to them with the compression ratio of the 350 being raised a little along with the power. Both models were fitted with a revised gearbox in which the gear pedal shaft was moved down to be concentric with the kickstarter and thus in a much better position. At the same time the clutch lifting mechanism was changed to a lever totally enclosed by the outer cover, while the neutral finder was retained. A dualseat was fitted as standard, a new air cleaner in an oval container and dual front brakes in a full width hub. The frame was modified in the area of the fork pivot, new cam forms were adopted to reduce noise and the competition machines sported Lucas wader magnetos.

Motor Cycling road tested a 500 cc Bullet in February and made it rather faster than the 'Blue Un' with a top speed over 85 mph. In general their findings were similar with the new double front brake improving that area and hauling the machine to a halt really quickly from any speed. To illustrate the detail care and attention they mentioned the siting of the rear brake light switch inside the left toolbox out of the weather. Not a great compliment to the wet weather performance of the switch but a sensible move to endeavour to preserve it.

In March came news of a quickly detachable rear wheel for the 500 Bullet which, while allowing speedy removal of the wheel, retained the usual cush drive in the hub, sprocket and brake drum assembly retained on the machine.

This was soon superseded when the 1956 models were announced as among the changes was one making the detachable hub a full-width

A flat-track Royal Enfield special with rigid frame as seen at Daytona in 1991.

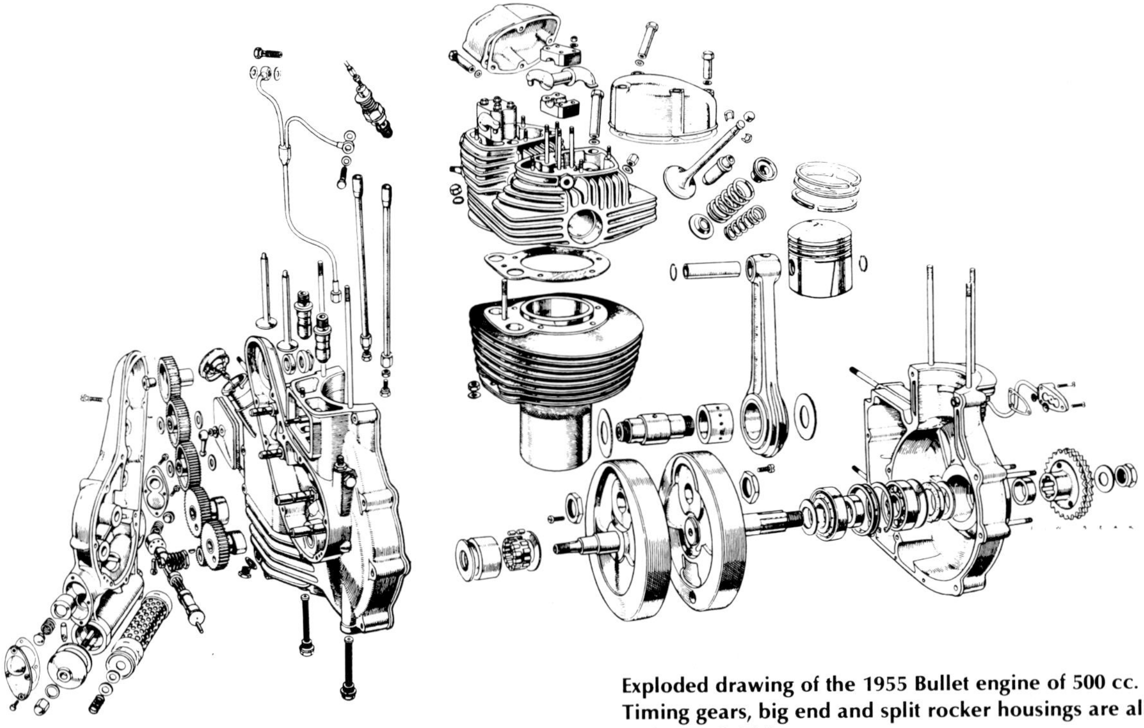

Exploded drawing of the 1955 Bullet engine of 500 cc. Timing gears, big end and split rocker housings are all typical Enfield features

The American Indian connection

Early in 1955, the firm announced a tie-up with Indian in the USA, as the result of which four models would be sold there, under the Indian label. Other than the name on the tank and raised bars, little else was altered with two singles and two twins on offer. The singles were the Woodsman, a 499 cc Bullet in trail trim with raised exhaust and suitable tyres, and the Fire Arrow, the 248 cc Clipper with the Indian emblem on its front mudguard. The twins were the Trailblazer, the 693 cc Meteor and the Tomahawk, the 495 cc Twin, neither much altered from their British form.

The link with Indian lasted until 1959, by when the range had stretched to ten models, all still heavily based on the stock machines and including the changes that had come along the way. Of the singles, the Woodsman was joined by the Westerner which was built for scrambles and fitted with a TT or GP carburettor and open exhaust. The Fire Arrow had been joined by the Hounds Arrow, a trail version, and both changed to the Crusader model as a basis.

The singles line was enlarged by offering two versions of the Ensign two-stroke in trail form. These were sold as the Lance in Mk1 and Mk2 forms, the first having more power from the 148 cc engine. The odd rear suspension remained while the finish was in Sahara yellow. A further single was the Patrol Car, a three-wheeled type popular in the USA and comprising a motorcycle front end with a framework carrying a large stowage box and supported by two rear wheels. The power unit for this was the 346 cc Bullet and the gear change was by hand.

Four twins were listed, the smallest the Tomahawk based on the Meteor Minor. The other twins used the 693 cc engine and comprised the Trailblazer, which lacked the Enfield casquette but was otherwise a stock road model, the Apache in trail-machine format, and the Chief which had an extended wheelbase and 16-inch wheels.

That 1959 season was the end of the connection, a few machines lingering in the showrooms into 1960. Once again, it demonstrated how flexible Royal Enfield would be to sell their machines.

Indian colours

1955

Trailblazer - Indian red tank, mudguards, toolbox and chainguard; black frame and forks.

Tomahawk - Indian red tank, toolboxes and air filters; polished aluminium mudguards; black frame and forks.

Woodsman - Indian red tank and chainguard; polished aluminium mudguards; black frame and forks.

Fire Arrow - Indian red tank, mudguards, toolboxes, air filter and chainguard; black frame and forks.

1959

Trailblazer and Chief - Indian metallic red or black; Apache - jet black; Tomahawk - metallic blue or black; Woodsman and Westerner - Indian metallic red; Fire Arrow - Indian red; Hounds Arrow - Chinese red; Lance - Sahara yellow.

In general, colours apply to tank, toolboxes, air filter and mudguards. Apache, Tomahawk, Woodsman, Westerner and Hounds Arrow have competition mudguards, either chrome-plated or polished aluminium. Frames in black.

ENFIELD CYC
AB573

light alloy one. Both models received a new diamond frame with no cradle under the engine but of all-welded construction. With this came matching side covers set in the rear frame loop, that on the right acting as the lid to the air cleaner box and that on the left for the tools and battery. The two boxes were formed in one to tidy up the centre area of the machine. Other cycle part improvements concerned the rear number plate, which was boxed in to the mudguard, and a new tank mounting. This comprised a bolt running across at the front into a steel sleeve set in a rubber bush in the frame and a spring steel sleeve set in a rubber bush in the frame and a spring steel clip at the rear to hold the tank down on rubber packing. The horn button was combined with the dipswitch on the left bar.

In the engines the 350 bottom half was made the same as the 500 by changing the width of the big end and making the timing side bearing a double row roller. The tappet feet were increased to reduce noise and wear while the 350 had a new cylinder head with larger inlet tract. Both machines had extensive electrical changes as the dynamo was dropped in favour of an alternator mounted on the left end of the crankshaft. This entailed revision to the primary chaincase and the wiring but left the ignition side alone, as this continued to be catered for by a magneto of the rotating magnet type. On the right side of the unit the kickstart gained a folding pedal.

The three varieties of competition Bullet continued to be listed but by 1957 had been reduced to one only, the scrambler. The road models were fitted with larger air filter elements and chrome plated tank panels, while the quickly detachable full width rear wheel became standard. The following year a new type of Burgess silencer was fitted and the rear tyre of the 500 cc model reduced in section to match the front and those on the 350.

The Dreamliner fairing built in 1956 in conjunction with *The Motor Cycle* and mounted on a 350 cc Bullet. Geoff Hay stands behind

During the year a new version of the trials Bullet made its appearance being based on the works model and fitted with alloy head and barrel, heavy flywheels, magneto and small bore Amal. It came with the full-width alloy rear hub with 7 in. brake and qd facility, while the exhaust system was tucked in even further with the tailpipe running inside the rear subframe. A small cylindrical toolbox was fitted under the seat and it used the old style gearbox with high mounted gear lever and kickstart curled round under the footrest. A front fork with reduced wheel spindle offset was fitted. The machine was a good example of a four-stroke trials model but the days of such were ending and the 350 Works Replica Trials Model, as it was known, was really too late for the average rider. Also it did not have the very much lighter hubs the works used or one or two other special items.

During 1958 Enfield had made the headlines by announcing a factory fairing for one of their 250 cc models and for the 1959 model line up followed this up with fairings for the whole of the range. Their interest in enclosure was not totally new for, in 1956, they had cooperated with *The Motor Cycle* to build the Dreamliner fairing onto an old 350 cc Bullet. This had been a rather futuristic experiment with full front and rear enclosure, widely spaced dual headlights and a deep screen. Tests had indicated a top speed improvement of 11 per cent and fuel consumption down by some 25 to 35 per cent, depending on the machine's speed. There had been no problems with the handling in strong, gusty winds.

Royal Enfield considered this type of full enclosure a little too radical for the general public so opted for a dolphin-style with exposed front wheel. With it came a very deeply valanced sprung front mudguard that ensured very full weather protection for any user. The new fairing was called the Airflow, this name being tacked

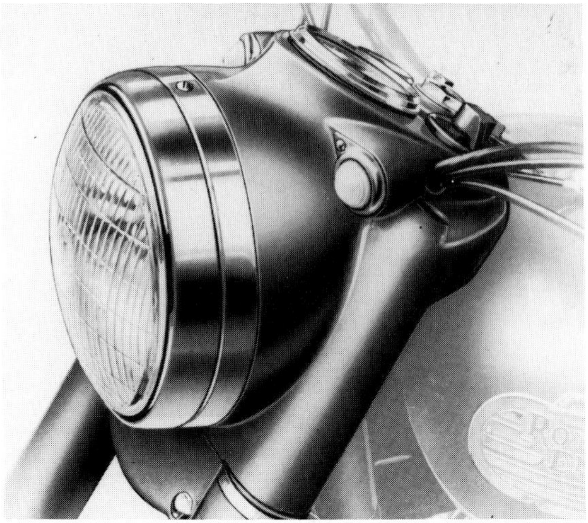

onto the model name where it was fitted.

This interest in enclosure was part of a trend at that period shown up by the Ariel Leader, bathtub Triumph and Norton machines and some European models. For a while it looked as if the public would follow this trend to combine motorcycling handling with scooter protection, but in a short space of time it was to reverse. Scooters were to fade from the scene and motorcycles went down the café-racer route to clip-ons and rear-sets.

As a further move to clean up motorcycling a rear chaincase was also offered as an option. There were a number of changes to the engines with the 350 receiving an increase in compression ratio, a bigger carburettor and new cams with more lift. That model was also fitted with 17 in. wheels, a 7 in. front brake and a larger petrol tank. The 500 also had its compression ratio raised and was fitted with a new alloy cylinder head with cast-in rocker boxes to carry the valve gear, each sealed with a cover held by a single nut. The larger Bullet retained its existing brakes and wheel sizes.

Both magazines ran road tests early in the year with *The Motor Cycle* having a 350 and *Motor Cycling* a 500. Both reported favourably, with the larger model logging 10 mph more than the smaller at 91 mph. There was a variance of opinion about the stands, possibly caused by the larger wheels fitted to the 500 making the task more difficult.

A further change had occurred on the 500 cc model when the test took place, with a move to coil ignition. The coil and ignition switch were both mounted in the right side toolbox that also carried the air filter element, while the contact points were mounted in a distributor unit assembly which replaced the magneto and so was driven from the timing gear in the same way as it. The 350 was altered in a similar fashion for 1960 when both models were fitted with a new style of casquette. This was common to nearly all the range and carried the twin parking lights further apart than before. In its top was housed the speedometer with the ammeter and light switch side-by-side behind it, the switch being on the left. The alloy casting ran back over the tops of the fork legs and encircled the steering head and damper knob when this was fitted. The 500 Bullet was also fitted with a larger petrol tank, a new shape of exhaust pipe and reverted to the dual 6 in. front brakes, having used the single 7 in. during the previous year.

The two Bullets, together with the trials 350, which became available to special order only, were reaching the end of their long lives for the days of the sporting single were numbered. In 1961 they were fitted with a new silencer which was common to the range. It was longer than the earlier ones with a tapered section leading to a parallel side Burgess-type unit fitted with a fishtail endcap which had the outlet directed down at the ground. The sections could be dismantled for cleaning.

This brought the models to their final year, 1962, during which they were dropped from the range after long and faithful service. The name, however, was to live on with a new unit construction 350, described in a later chapter.

Far left **The famous casquette and the infamous twin pilot lights never to be used when on the môve. Neat instrument panel**

Left **For America—the 1957 Indian West Coast Westerner, otherwise the 500 cc scrambles Bullet to Limeys. One of several models sold under the Indian label in the USA**

Below **Late type 500 cc road Bullet with coil ignition and revised silencer**

4 Twin

After the second world war there was a great rush among the British firms to get a twin cylinder model into their lists in opposition to the Triumph which had come out first in 1937. Not for nothing was the Speed Twin advertised as the one with the 10 year start.

In time all the major firms produced their twins as fast as the postwar problems would allow them and Royal Enfield were no exception. It was late in 1948 before it could be announced and, while it followed the accepted pattern of the English vertical twin, it did have a good many design features that were unique at the time.

While Enfields were not afraid to design and market their own ideas as to how a vertical twin should be built they also used many of their existing practices where these had proven successful over the years. Thus the oil was carried in a sump cast-in with the crankcase behind the flywheel compartment and the oil pump was of the double acting twin plunger type located in the timing cover, both features copying the Bullet layout. Different, however, was the position of the main oil filter which lay across the engine in a separate tunnel cast-in the crankcase just below the Bullet style oil filler. In addition the filter was on the return side of the scavenge pump and worked on the oil as it went back to the sump.

The engine was based on dimensions of 64 × 77 mm to give it a capacity of 495 cc and, as was usual at the time, had the crankcase split on the vertical centre line. Each barrel and head was separate from the other which improved access but reduced the overall rigidity of the engine. The crankshaft was a one-piece alloy iron casting formed with a substantial central flywheel. It had small outer flywheels as well and the crankpins were cored out to allow oil to pass through. This passage was sealed with steel discs retained by circlips and the one on the right carried a spring-loaded oil release ball-valve. Cross holes fed the crankpin bearing areas. The left mainshaft was hollow and had a non-return disc valve fitted to it which communicated with an external pipe and acted as the crankcase breather.

Both main bearings were the same size with a ball race fitted on the left and a roller on the right timing side. The big ends were plain and the alloy rods ran directly on the crankpins. The caps were attached by fitted bolts secured by nuts and split pins. This again was unusual but to be expected in view of Enfields long experience with the plain bush big end on the singles. At that time they

The production 500 Twin in January 1949. The model never had any other name while in this form

The prototype 500 cc Twin in March 1948 with ammeter and switches on side of box. Many cycle parts common with Bullet

were not alone in fitting throw-away rods, for Triumph did the same although others preferred the insurance of shells. The gudgeon pins also ran directly in the rods and were held in the pistons by circlips. Each piston was slightly domed to give a compression ratio of 6·5:1 and carried a scraper and two plain rings.

The pistons ran in cast-iron barrels which were spigoted into the crankcase for over half their length. The two were interchangeable and were spigoted up into the light alloy cylinder heads, both items being held down by five through studs, three along the outer edge and two on the inner. Each head was cast with two integral rocker boxes and had cast-iron inserts for the valve seats and sparking plug plus a cast-in steel exhaust port stub. An inlet manifold was bolted to the rear faces of the heads to carry the single Amal carburettor.

The valves worked in pressed-in guides and each was held to its seat by duplex valve springs retained by a collar and split cotters. Hardened valve caps were used. Each rocker was free to oscillate on its spindle which in turn could float in its inboard bearing machined in the rocker box and the outer one formed in a detachable plug. Valve clearance was adjusted with a screw in the outer end of each rocker arm locked in place by a nut fitted under the arm. This nut had a conical seating and a shallow saw cut across it so was self-locking as it was tightened. All rocker spindles were hollow and supplied with oil from the secondary side of the pressure pump via drillings in the crankcase, barrels and heads. Thus there were no external oil pipes at all which gave the engine a very neat style. The rocker boxes were closed with separate plates each held by four nuts.

The rockers were moved by push rods situated in the four outer corners of the engine and this was done to allow the maximum flow of air between the cylinders. Each rod was the same length and in solid steel with forged cups at each end. Flat base tappets ran in iron guides pressed into the top face of the crankcase and the camshafts were two in number mounted fore and aft of the cylinders high up in the crankcase. They ran on bushes and the cams were widely separated on the shafts but very close to the supporting bearings. Both inlet and exhaust cams had quietening ramps.

The drive to the camshafts was by a single strand chain as the firm considered this better than a gear train when high mounted camshafts were used. The crankshaft sprocket was machined with a male taper which keyed into the

Left side of prototype Twin. The heavy sprung front mudguard spoilt the lithe lines of the rest of the machine

hollow mainshaft and was held by a special bolt. This bolt screwed into a knurled nut which sat in a recess in the back of the right flywheel. To prevent it rotating and to hold it in position for assembly it was locked by a cross pin dropped into a radial hole in the flywheel and this was held by a grub screw itself locked by centre punching. The special bolt also did another duty as its head was machined with the worm drive to the oil pump shaft.

Each camshaft carried a keyed sprocket on a taper and the chain was adjusted for tension by an eccentrically mounted jockey sprocket situated between the inlet and crankshaft ones. The adjustment was done by moving a quadrant pressing through an arc, this part locating to the eccentric spindle by two flats. Clever detail design allowed the quadrant plate and the spindle to be reversed if needed to provide the maximum possible adjustment that the spindle offset was capable of giving. Simple markings on the sprockets and the keyways in the shafts made re-timing easy.

The inlet sprocket had a second row of teeth machined on it and this drove a dynamo mounted behind the cylinders which sat in a small cradle bolted to a platform formed on the upper surface of the crankcase. The dynamo was strapped down onto this cradle and ran at engine speed. It was a special type with an extension at the driven end containing skew gears which drove a vertical shaft at a 2:1 reduction. This ran in a distributor unit mounted on the dynamo which contained the points, advance mechanism and distributor to feed the high tension to the two plugs. A separate ignition coil was used and the whole distributor unit could be rotated in its housing to adjust the ignition timing.

Left **The production Twin in November 1948**

Right **The 500 Twin engine without the gearbox in place bolted to the rear crankcase wall. Note special dynamo with skew gear drive to points and distributor housing**

Below **The 500 Twin in 1953. Speedometer housed in fork crown and switch plus ammeter in headlamp, which has underslung pilot of very doubtful use**

The whole of the timing side, including the dynamo drive, was enclosed by a single cover which contained the oil pump, the quill feed to the crankshaft, and support for the eccentric sprocket spindle. The lubrication system was similar to that of the other Enfields with two supplies, one to the crankshaft and the other to the rockers. The first of these dealt with the big ends and their overflow splash lubricated the cylinders, pistons and mains. The drain from the rockers passed down the push rod holes to the camshaft tunnels which were cast into the crankcase halves to isolate them from the flywheels. Some oil in the inlet side was fed to the rear of the cylinder walls but in both tunnels oil was allowed to rise at a set level, after which it overflowed into the timing case. From there it was picked up by the scavenge pump and returned to the tank via the main filter. The other side of the pump collected the oil from the sump and returned it along the same route.

As on the Bullet models, the gearbox was bolted straight to the back of the crankcase and driven by a slipper tensioned duplex primary chain. It and the four-plate clutch were enclosed in a cast aluminium chaincase, as on the single, and the clutch was operated by a quick thread worm.

For the rest, the twin was a repeat of the Bullet with the same layout of frame, swinging fork rear suspension, telescopic front forks, hubs, wheels and general equipment. The last included two toolboxes, a saddle, sprung front mudguard, detachable rear guard and an air cleaner box that was sandwiched between the battery carrier and a further box carrying the ignition and lights switch and an ammeter. Thus the headlamp shell was bare of controls and the speedometer was fork mounted.

The exhaust pipes were fixed to the port stubs with finned clamps and swept down to a silencer on each side mounted parallel to the ground. The finish was polychromatic silver grey for all the painted parts, except the air cleaner box, headlamp shell and ignition switch box which were black and the petrol tank. This last was chrome-plated with frosted aluminium panels lined in red and with the maker's name in the same colour. 'Royal Enfield' was also engraved into the timing cover and the whole machine looked very smart and stylish.

It was listed at just under £200 but very few machines reached the home market as nearly all the production was taken for export. For 1950 it gained the same front end changes as the Bullet with the cast alloy fascia carrying the speedometer, the fork modifications and the shallower front mudguard. In addition, the combined switches for the electrics moved onto the front face of the switch box leaving the ammeter facing outwards and completely unreadable to a mounted rider. The changes for 1951 were also apparently minimal, with the adoption of the forged fork ends and unsprung front mudguard. Less obvious was a change to die-cast crankcase halves which emphasised Enfield's belief that their design was correct. This move also dropped the weight to 390 lb.

At the end of 1950 both magazines had a chance of a short ride on the twin and reported well on it. A little later *The Motor Cycle* conducted a formal road test and confirmed easy starting, good handling and a nice engine performance. They managed to get the valves floating in top gear at 85 mph, which perhaps indicated a need for another tooth on the gearbox sprocket, but found the brakes were both spongy, the petrol tank too wide for optimum comfort, and both stands left a good deal to be desired. Brakes aside, however, the twin was definitely a high performer. A few weeks later *Motor Cycling* confirmed these findings and found one mph more.

For 1952 there was nothing really major changed but a number of useful detail modifications were carried out. A noticeable one on the engine was a move to external pipes for rocker lubrication, these running from the crankcase to

Indian Trailblazer 693 cc Twin of 1955. American logo is matched by handlebars, tyres and front fender lamp

A late 1956 picture of a Meteor Minor engine in a Crusader frame. A tight fit but it went in

the outer ends of the spindles. Inside the engine two more screws held the crankcase halves together in the region of the barrels. The electrics were rearranged with the ammeter and light switch moving to the headlamp, and the ignition switch to the top corner of the right toolbox. In place of the old switch box was fitted a larger air cleaner and this change made access to the filter element very much easier. As on the Bullet, the breather gained a banjo union on the chaincase and a dualseat became an option available for new machines.

In 1953 the 500 twin began to be overshadowed by a new 700 cc version but its presence in the programme was emphasised by a change to die-cast cylinder heads. This represented a considerable tooling investment but one well worth taking on an established design. The new heads had the valve seats shrunk in and were fitted with a new Y-shaped inlet manifold in place of the earlier type. Otherwise the changes to the machine were as on the Bullet and concerned the underslung pilot lamp, stop and tail lamp, reflector and floating brake cam spindle housings.

The following year the small twin was fitted with the casquette and proprietary rear suspension units and was offered with the option of the existing coil and distributor ignition system or a mag-dyno. If the latter was fitted the magneto sat in the normal dynamo position and carried the dynamo on its back. A curious step as other models were being fitted with alternators and rectified electric systems. In order that the magneto ran at half engine speed the size of the second of the two sprockets attached to the inlet

camshaft was halved. The dynamo speed was recovered by the gearing up between the two electrical instruments.

It was a model equipped with the mag-dyno that *Motor Cycling* road tested early in 1954 and they confirmed earlier findings with regard to the ability of the twin to put the miles behind it quickly and smoothly. Once again the brakes were considered spongy and the engine breather oil leak onto the rear chain was totally unable to keep it lubricated. The main light beam lacked spread for winding roads, but otherwise the machine received good marks.

The braking problem was dealt with in the 1955 programme with the adoption of the dual front brake. At the same time the mag-dyno became standard and the new gearbox end cover, with re-positioned gear pedal, was adopted as on the Bullets. Frame changes and the oval air cleaner body also copied the singles.

During the early part of the year the twin became available with the option of the quickly detachable rear wheel and, as with the singles, this was superceded by a full width version for the 1956 models. At the same time the compression ratio rose to 7·5:1 and the power to 27 bhp at 6000 rpm, while the tank mounting, folding kickstart lever, and combined horn button and dipswitch were all fitted.

1957 brought some more radical changes with the dynamo being replaced by a crankshaft mounted alternator. This allowed reversion to coil ignition with a distributor unit mounted in the old magneto position and driven by chain from the camshaft. This in turn left room for a Monobloc carburettor to be fitted in place of the old type with separate float chamber, this not having been possible before due to the space restriction. The twin was further altered with the engine being fitted into the frame first used by the Bullets the previous year and with a slight lowering of the gearing. The appearance was enhanced with a modified fixing for the rear mudguard employing swept-back stays on each side which blended into the structure cleanly. For 1958 all the twin received was the new style of silencer, for the very good reason that a major redesign was to be launched that year.

The new 500 twin appeared in April 1958 and was given a new name—the Meteor Minor—and listed in standard and de-luxe forms. It followed the same basic design as its predecessor but represented the changes in technology of the previous decade with new engine dimensions.

Left **1957 Meteor Minor 500 with siamezed exhaust pipes. Fitted with saddle only despite the pillion footrests**

Right **An interesting March 1957 picture of a twin fitted with the leading link forks later to appear on the Super 5**

Right **The 1961 version of the Meteor Minor Sports. Siamezed exhausts and provision for rev-counter drive on timing cover**

Far right **The de-luxe Meteor Minor for 1961 with few external differences from the Sports model**

Below **Line drawing of the 500 Twin engine showing clearly the pre-war ancestry of the timing chain layout. The magneto, when fitted, ran at camshaft speed**

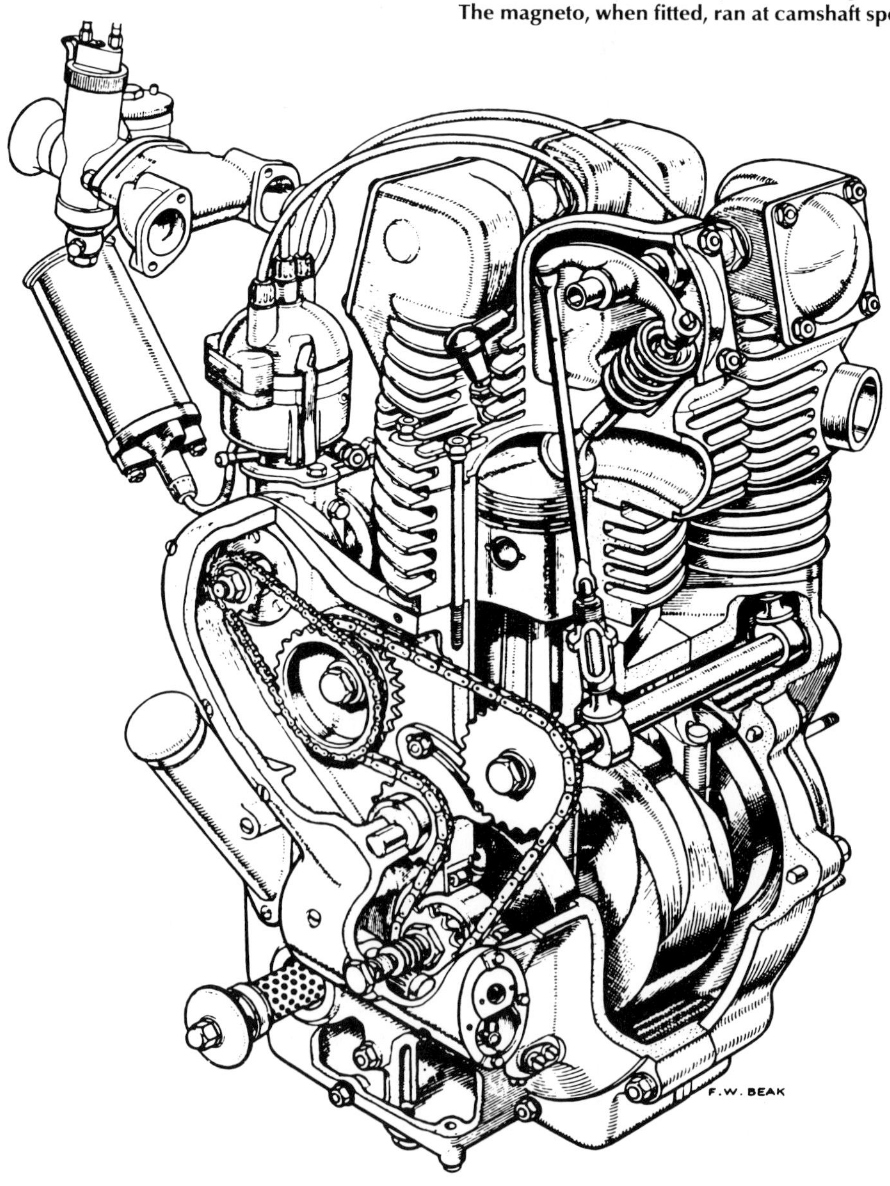

These were 70 × 64·5 mm and the short stroke allowed the use of bigger valves while making the pistons common with those used in other models. Compression ratio was up to 8·0:1 and power to 30 bhp at 6250 rpm. The construction of the engine followed the established Enfield pattern in most ways with the push rods at the corners of the barrels, separate parts above the crankcase mouth, chain driven high camshafts, alternator, and distributor assembly behind the cylinders. Oil was still carried in the crankcase and the four-speed gearbox bolted to its back wall.

Different were the big end bearings which became split shells running on a one-piece nodular iron crankshaft which itself turned in ball and roller bearings, as always. Alloy rods were used along with alloy heads and iron barrels. The shape of the timing cover was altered so that it had the look of a pressing but it still housed the oil pump and also the filter, as on the singles.

The primary drive was by duplex chain as before and drove a multiplate clutch. This differed from other Enfield clutches in that, although the plates were held together by six springs, these were in turn held by an outer plate bolted to the clutch hub instead of the more usual spring cups. This outer plate carried an angular ball race and through this passed a shaft which extended behind a second race fitted to the pressure plate. Pull on this shaft lifted the plate and separated the clutch and was obtained by a ball and ramp mechanism outboard of the clutch on the left. This contained four balls which rode up a ramp as the clutch lever was moved, and so pulled on the through shaft. The clutch cable ran up and forward inside the primary chaincase to emerge from its upper surface and sweep up to the handlebars.

The gearbox was the usual Albion complete with neutral selector and kickstarter with folding lever mounted concentric with the gear lever. The complete power unit went into a simple frame very similar to one already in use and equipped with swinging fork suspension at the rear and the Enfield telescopics at the front complete with casquette. It was fitted with a siamezed exhaust system, the silencer being fitted on the right low down and 17 in. wheels which helped to make it appear a very compact machine. At the rear a 7 in. brake was fitted and, on the standard model, a 6 in. full width one at the front. The de-luxe version received a 7 in. front brake and a quickly detachable rear wheel which was definitely needed as it also had full rear chain enclosure.

The standard model was fitted with a saddle but the de-luxe had a dualseat and pillion rests, an air cleaner, stop light, prop stand and a well valanced front mudguard which had panels to carry the registration number on the sides. The de-luxe version also had a pressing to cowl in the distributor assembly. The centre section was designed as a unit with the air cleaner in the centre and boxes on each side with their own lids. That on the right housed the ignition coil and the tool roll, while the one on the left had the ignition switch built into it. Light switch, ammeter, and speedometer continued to live in the casquette along with the two useless pilot lamps.

Finish was in black and chrome on the standard model, but there was the choice of three two-tone options on the de-luxe with burgundy, blue, or silver grey for the tank, mudguards and toolbox lids, while the frame, forks and centre section remained black.

Smart, but to some eyes the new lines had lost some of the light, lithe appeal of the older twin for the line of the new timing cover, the casquette, the small wheels and the dualseat all went together to produce an impression of a small but heavy machine. In fact it was 20 lb lighter and geared up so went faster but, somehow, lacked the style of its predecessor.

The Motor Cycle road tested a de-luxe version of the revised twin in June 1958 so had the benefit of the larger front brake and enclosed rear chain. Less popular were footrests too far forward for high speed cruising and the top edges of the seat pan which dug into the rider's thighs. The low seating position given by the small wheels also meant the footrests had to be rather close to the ground and so could be scraped on it despite the firm suspension.

The performance was up to 89 mph with a high of 97 with a strong following wind. Acceleration was brisk, with the new clutch taking the drive up smoothly and gear changes quick and clean going up. Coming down the engagement of the dogs could be felt and, on occasion, the clutch plates tended to stick so bottom would not engage at rest without a jerk. The brakes worked well and the machine was stable but inclined to weave on fast bends with undulating surfaces.

Some vibration was apparent at the higher speeds, while the machine noise level was of an acceptable standard. Maintenance was to the usual easy Enfield standard with single bolts holding tank or chaincase cover in place. All told, an acceptable motorcycle that reflected the call for machines with a lively performance. However, for many the charm of the 500 twin had gone for that was part appearance and part the soft power delivery that gave such restful cruising.

In July the Meteor Minor was added to the list of machines that could be fitted with the Airflow fairing and this facility continued to be available for all versions of the 500 twin from then on. When fitted it was accompanied by the very large sprung front mudguard that went with it. During 1959 the standard machines, with or without the Airflow, were dropped but in their place came a Meteor Minor Sports. This was based on the de-luxe Minor but had special cams, valves and springs to raise the power a little. It was fitted with chrome plated mudguards and tank, the latter with its top painted with a full width stripe. The rear tyre increased its section a trifle.

For 1961 the de-luxe model was fitted with the new silencer as used on the Bullet, still with a siamezed system. Both models had the re-styled casquette but the price of the Sports model was held down by keeping the specification and finish simple. Thus, although it had the more powerful engine, it also had the older form of silencer, lacked the qd rear wheel and was vapour blasted to give a satin finish to hubs and timing, chaincase and gearbox covers. Its paint finish was a two-tone in mist grey and black, while the de-luxe was more colourful in burgundy red or peacock blue.

A rather special Meteor Minor was seen that year in the Isle of Man during TT week. It was a development machine not a 1962 prototype and was based on the de-luxe model fitted with an Airflow. This itself was slightly modified with indicators built into the moulding either side of and below the headlight. Further fibreglass was used to form a very neat distributor cover and the machine had a new fork crown piece with an Airflow nameplate. At the rear the machine was enclosed rather in the style of the Ariel Leader but by a one-piece moulding that incorporated panniers on each side. In front of these, parts of the moulding could be released at the turn of a single screw to give access to the battery and the electrics. At the rear a door hinged down to give access to both panniers and this carried the rear number plate and light. On top of the enclosure behind the seat was bolted a large carrier.

It created a lot of interest in the Island but was not in the 1962 range, which continued with the two models, the Sports fitted with the dual front brake and available in burgundy or blue, and the de-luxe in a two-tone burgundy and cream finish. *Motor Cycling* road tested a de-luxe model with Airflow and panniers in May and found it just the ticket for touring at cruising speeds up to 70 mph. The suspension was a little on the hard side but that aside it proved to be a very comfortable machine and ideally suited to its purpose.

A little sadly, this was not what sold motorcycles at that time for a decline had set in and was to last a decade, destroying much of the British industry in the period. The de-luxe version of the Meteor Minor was an early victim and vanished from the lists in 1962. At the same time the more sporting model became called the 500 Sports Twin and reverted to the 7 in. front brake, although it retained the siamezed pipes. Twin, separate, exhaust systems were available as an option.

The last of the line had a new finish in gold for the mudguards, box lids and tank top but this could not save it in a motorcycle world split by the needs of learner 250s and high performance. There was no longer any place for a 500 twin and so, in 1963, the final example of the small Enfield twin went from the range.

Last of the line, the 1963 500 Sports Twin. Note alloy centre stand

5 Big Twin

If a twin was good, then a bigger twin had to be better was the thinking in the early 1950s, encouraged by the US market that called for sheer 'cubes' to cope with long straight roads that demanded the ability to cruise fast. This could be done with a hot 500 but was much more restful if a larger machine was used. In 1949 if you bought British that meant the Ariel Square Four or the Vincent vee twin, both rather expensive and not everyone's 'cup of tea'. For the other factories the much easier answer was simply to open up their existing vertical twins and this quickly produced the BSA Golden Flash and Triumph Thunderbird.

Next to come was the Enfield to be followed by others, but for many years the Redditch machine was able to claim that it was England's largest vertical twin. It was of 700 cc and based on the 70 × 90 mm dimensions of the 350 cc Bullet. The size was not arrived at just by doubling the capacity of the single, but from a wish to use the crankcase castings of the 500 twin. The physical dimensions, strength and rigidity allowed the design team to go past the 650 cc size chosen by BSA and Triumph to the full 700 cc and this in turn led to the Bullet sizes. Thus the experience of the single in combustion chamber layout could be transferred straight over to the big twin.

The Meteor engine was thus an amalgamation of the 500 twin crankcase and the 350 Bullet top half, this being redrawn into the format of the twin. Externally the cylinder heads had an extra fin and all the fins on the heads were longer. Inside, the restriction of the crankcase size prevented the big end cap being assembled as on the 500. The rod was still in alloy but fitted with shell bearings and the cap was retained by some very special socket head cap screws. These were ground to locate the cap to the rod and screwed directly up into it. At the small end the gudgeon pins still ran directly in the rods.

Compression ratio was 6·5:1 and the engine pumped out 36 bhp at 6000 rpm, which meant an extra plate had to be fitted into the clutch. The chassis also had to be strengthened and this was done by stiffening up both suspension systems, modifying the internals of the damping and stiffening the rear fork assembly as was also done for the 500 cc Bullet. To cope with the higher weight and speed the 6 in. dual front brake with its whiffle tree compensator at the lever was fitted. At the rear went the 7 in. unit.

The general equipment was as on the 500 twin with deeply valanced mudguards but the fuel tank was larger at 4 gallons and the dualseat was still an option. Tyre sizes were as the smaller twin and the machine was finished in polychromatic copper beech, including the headlight, with its underslung pilot lamp, and the air cleaner box.

The 1953 Meteor 700 cc engine showing its close affinity to the smaller twin. Under the skin were two 350 Bullets

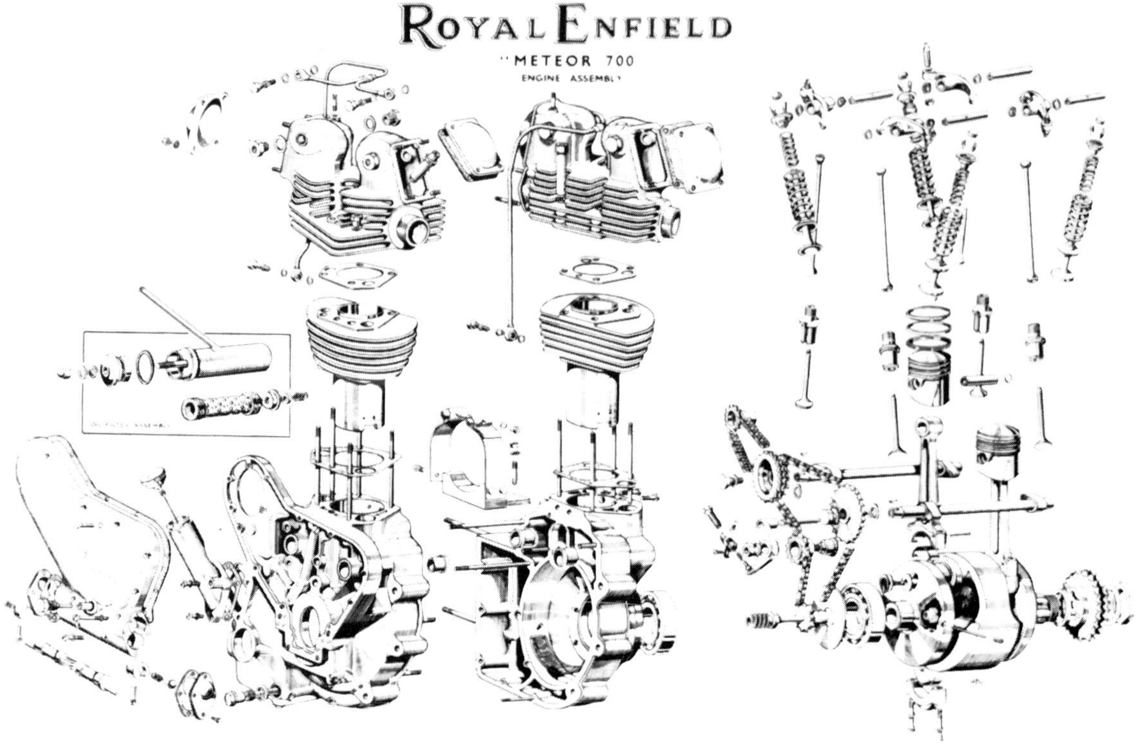

The dualseat, when fitted, was in a matching colour. The machine was also available with the sidecar forks, gearing and a steering damper, while further options of legshields and panniers could be had.

Early in 1953 *The Motor Cycle* had a run out on the new model and reported that the handling was first class, the performance potential definitely lurking in the still tight engine, but the front brake not as powerful as it should have been. The machine was very tractable and the only vibration recorded occurred when the head steady worked loose. It disappeared once the stay was re-tightened. Later in the year *Motor Cycling* carried out a formal test, recording a maximum of 94 mph. The machine would also run down to 20 mph in top gear and accelerate away without pinking or snatch.

The big engine proved so smooth and re-

Above **An exploded drawing of the Meteor 700 engine showing the details of construction. Same oil pump type, different filter position**

Right above **The Meteor 700 as first seen in 1953. Fitted with the optional pannier in its very sturdy frame**

Right **A March 1954 picture of a 700 fitted out for Police use. The rear box is the massive radio of the day, now an electronic chip**

RE 700

Left **Very nice Meteor and Watsonian chair on show at Earls Court in 1955**

Right below **The 1955 Indian Trailblazer 700 giving the American customer the 'cubes' he wanted**

sponsive that there was never any need to hang onto the gears and the riders found themselves changing-up early and letting the engine pull. Cruising speeds were in the 70 to 80 mph region without fuss. There were indications in the test that both clutch and gearbox were working hard for the box was stiff to operate on occasion and the clutch lever had to be pulled right back to the bar to ensure it freed the clutch completely. It was, however, very light in operation.

The braking figures recorded were good but the front needed very firm pressure on the lever to achieve this. The magazine suggested this was due to the friction of the two cables and the added load of the extra pull-off springs in the second brake. It was more likely, in view of the comments made in tests of other Enfields, that the brake just was not large enough and that a move to dual 7 in. or a single 8 in. was what was really needed but was never to appear. That aspect aside, it was a fine motorcycle capable of carrying its rider far and fast.

There were few changes for 1954 but the Meteor, like most other models, was fitted with the casquette and proprietary rear suspension units. With the adoption of the casquette it was no longer easy to change the fork trail by fitting different crowns. The solution was a variety of lower forks with longer lugs to give reduced trail for sidecar use. One further small change was to the front mudguard whose lower stay was run outside it so it could double as a front stand when needed.

1955 brought the revised gearbox end cover and re-positioned lever as used on the other models, the modified frame, and the oval air box. Less obvious to the eye was an increase in compression ratio to 7·25:1, and of power to

39 bhp at 6000 rpm. As on the 500 twin, the electrics were changed to a mag-dyno mounted in the old generator position.

In January *The Motor Cycle* tested a Meteor to sidecar specification with a Canterbury chair hitched. It highlighted the ability of the large engine to haul a large child-adult saloon along at a 60 mph cruising speed indefinitely. With the lowered gearing the machine was free of snatch down to 12 mph in top gear and, by judicious use of throttle and ignition advance, would pull smoothly all the way up to 70 mph. The outfit was equipped with a sidecar wheel brake and this enabled it to stop in a reasonable distance, although this was more than the solo figure. There were a few detail snags but in all the machine was enjoyed.

As on the 500 twin, the quickly detachable rear wheel option was made available for the Meteor during the year but was soon superseded by the 7 in., full width, light alloy version.

When the range for 1956 was announced the Meteor was seen to be replaced by a new and extensively redesigned version called the Super Meteor. The engine looked much as before but the power was up a little to 40 bhp, while the equivalent engine speed was reduced to 5500 rpm. The compression ratio remained at 7·25:1 but the inlet valves were enlarged and the porting worked on to produce the improvement. The engine was fitted with a new crankshaft with larger crankpins and the outer webs cut-away. Both heads and barrels had deeper finning and the head spigot locating them deleted. In its place were fitted two hollow dowels around the outer fixings and this change enabled the head gasket to be improved. The push rods were changed in material to light alloy tubes with hardened ends.

Lubrication was improved by fitting a two start worm to double the speed of the oil pump and by arranging for the felt filter to be in the supply line from pump to big end bearings. Ignition was still by magneto but the electrical system was powered by an alternator attached to the end of the crankshaft in the primary chaincase. With the old mag-dyno out of the way the carburettor became a Monobloc. Gearing was raised a little,

Royal Enfield
Constellation
for sheer performance
700 c.c.
50 b.h.p.
CON 700
ROYAL ENFIELD
Constellation
Retail Price £236.9.6
Purchase Tax £58.10.6
£295.0.0

RE 700

Above **A Constellation hitched to a single seat sidecar to make a fast sporting outfit. Note oil stains on exhaust and sidecar trail fork legs**

Left above **A 1958 brochure for the Constellation, or Connie as it became known to most owners**

Left **The 1958 Super Meteor with casquette and twin exhaust systems**

and a folding kickstart lever fitted.

The engine unit was fitted into a new frame, also used for the Bullets, with single top and front down tubes brazed into the malleable steering head lug. Loops ran round from the top tube under the gearbox and supported the tops of the suspension units and the swinging fork pivot. In the space under the nose of the dualseat went a divided box with lids on each side. The left side contained the air filter and tool roll and the right the battery. As on some other models, the rear number plate was boxed in and the horn button and dipswitch combined. The fuel tank mounting was modified as on the 500 twin.

At last Enfields had a 100 mph machine, for on test *Motor Cycling* managed to exceed the magic ton by one mph. That the Super had not lost the tractable nature of the early Meteor was shown up by an overall consumption figure of 55 mpg which rose to 70 on one long run down from Scotland when 75 mph was the maximum speed used on the normal roads then available. In this test the brakes proved to be up to the mark and, in most other respects, it was a repeat of earlier findings. Possibly the low speed pulling had suffered and the engine no longer ran to such low speeds in top without some mild protest but this was never a problem. One aspect that did receive criticism was the light alloy centre stand for it took great strength to raise the machine onto it. Unfortunately the prop stand was barely man enough for the weight of the big twin as it remained the typical Enfield spike of the past. One other snag concerned the fit of the toolbox but all these irritants were forgotten on the road when the benefits of the large capacity came into their own.

For the next two years the 700 was produced with few alterations, for in 1957 only the rear mudguard support was changed as on the 500 and an improved magneto fitted to give better starting and slow running. The next year this was displaced by coil ignition as already in use on the smaller twin with the distributor housing in the old magneto position. At the same time the silencers lost their tailpipes to give them a neater line.

In April 1958 the Meteor Minors were announced and with them came an even higher performance 700, the Constellation, quickly nicknamed the 'Connie'. This had an engine unit based on the Super Meteor unit fitted with pistons giving an 8·5:1 ratio, magneto, TT carburettor and siamezed exhaust system. This pushed the power up to 51 bhp at 6250 rpm so the gearing was lowered a little and the Meteor Minor type clutch fitted. As on the new 500 twin, a new frame was employed based on one already in use but increased in size to accommodate the 700 and its power. The chassis specification remained very much as on the Super Meteor and included an air filter and a handsome $4\frac{1}{4}$ gallon petrol tank. A steering damper was fitted as standard. Finish for the tank was chrome-plating with the top painted polychromatic burgundy as a broad stripe. The mudguards were also chrome-plated, while the centre section lids matched the tank stripe and the frame and forks were in black.

And it went too! *Motor Cycling* took one to Belgium and timed it at 115 mph, also averaging over 80 from Brussels to Ostend. Under all circumstances it was a very quick machine with that ability to gobble-up slower traffic without effort that only seems to come from capacity. It recorded 51 mpg which was very good under the hammering it took and also proved to be perfectly docile in London traffic. Only at one point did it let itself down as, on the approach to Ostend at around 110, one of the rocker oil feed pipes parted from its union and deposited oil over rather too much engine. The only other snags concerned oil mist around the breather, flooding of the carburettor if the prop stand was used, and the inability of said stand to keep the machine upright unless it was also put into gear. But it did go well!

At the end of the year the Airflow fairing became available for both the 700 cc models and the Super Meteor was fitted with the siamezed exhaust system like the Constellation, unless the fairing was required. In that case the normal twin pipes and silencers were retained.

In 1960 the Constellation was fitted with twin carburettors, Amal Monoblocs each carried on a short inlet stub so that the carburettors flared out enough to clear one another. Even so the right unit had to have its float chamber compartment chopped off and fitted with a blanking plate in order to get it into the available space. It was supplied with fuel from the standard left carburettor by a short flexible pipe joining the two main jet holders. The centre section was also modified with small ears on the lids to shroud the intakes and depressions in the boxes to clear the carbs.

The Constellation in 1959 fitted with the factory Airflow fairing and its complementary and massive front mudguard

The engine was altered internally and by this time all crankshafts had a higher balance factor and a bobweight shaped flywheel. Each Constellation crankshaft was dynamically balanced and had an oil breather in the driveside mainshaft as well as that bolted high up on the left crankcase outer wall. The mainshaft clack valve breathed into the primary chaincase and this in turn was vented to the outside world. A magnet was fitted in the main filter as a precaution.

The following year, 1961, brought more changes with the compression ratio of the Constellation down to 8:1 and the three-part silencer as used on the 500 twin fitted to both 700s. Both were fitted with an extra plate in the clutch and this was lifted by a push rod through the mainshaft which replaced the earlier ball and ramp mechanism. Both continued with siamezed exhaust systems unless fited with the Airflow fairing, and were fitted with a new type of battery and a revised fuel tank mounting. Both models were fitted with a neat rear enclosure assembly that combined the seat, its subframe, rear mudguard, a lifting handle on each side and tyre pump location with a fibreglass moulding. This formed a deeply valanced rear mudguard to follow the lines of the seat and blend with steel pressings on each side. These pressings followed the line of the centre boxes and rear subframe to match the moulding and give some enclosure, while retaining the motorcycle look. The Constellation was fitted with low, sports handlebars and the Super Meteor received the Connie's fuel

Cutaway line drawing of the 1961 model Constellation. Twin carburettors, magneto, some rear panelling—very fast

AMAL
COLLINS

tank but painted, not plated, and was geared down to the same level.

The new fibreglass moulding did not last long and by January was replaced by steel pressings with the same appearance. At the same time the rear fork bearings were changed from bronze to Silentbloc bushes. A formal road test confirmed the earlier findings, although the electronic beam at MIRA gave a top speed just over 112 mph and showed that the hand timing over a Belgium quarter-mile had been a little optimistic, but only by a fifth of a second. Comfort and handling were good and the brakes worked well from high speeds.

They proved to be less effective in another test lacking bite and needing considerable pressure, a sure indication that they were just too small for their job. There were other detail faults with the clutch which was marginal unless cared for, and the horn button which was out of reach, but this was not over important as the horn note was itself feeble. One or two oil leaks occurred and these became more of a flood when some timing cover screws vibrated out. Fully offsetting these minor irritations was the performance which brought out descriptive words such as 'mile-gobbler' and 'hog-bus'.

1962 Super Meteor with rear mudguard styled to match side panels, seat and toolbox lids. Later type casquette and silencers

There were no mechanical changes for 1962 but the finish of the Super Meteor became two-tone in burgundy and cream. The Constellation retained its chrome-plated tank with broad top stripe and remained the sports model, while the Meteor was aimed more for the tourer and sidecar man.

It was in this guise, with a Blacknell single-seat sports chair attached, that *Motor Cycling* road-tested one in their very last issue as a magazine, in February. With a lower compression ratio than the Constellation it proved to be very suited to its task on ice and snow covered roads. When conditions were less slippery and the ample power did not try to spin the outfit it could be cruised along at 65 mph. The brakes failed to record a very impressive figure in the test panel but, in practice, worked well on the road to augment the fine handling of the well lined-up outfit.

Late in the year the programme for 1963 was announced and both the 700s went to be replaced by a sidecar only Constellation and the larger capacity Interceptor, described in the next chapter. The new version of the Constellation was an amalgam of the two earlier machines and used the Meteor engine with the compression ratio down to 7:1, a single Amal carburettor, and coil ignition to produce 40 bhp at 5500 rpm. A siamezed exhaust system was fitted as standard, although twin pipes were available as an option and fitted when the Airflow was in use. The machine had all the Enfield sidecar features with lowered gearing, reduced fork trail, stiffer suspension front and rear, steering damper and, in addition, was fitted with the same size tyres front and rear, both with square section for sidecar use. The machine was finished in flame and cream and was also available in solo trim to special order.

Not for long, however, for the market for sidecar machines was very small and the accent was on capacity and power. Thus it was to their largest twin that the company turned and the big twin dropped from the lists after over a decade of service.

There was, however, one other aspect of the life of the big twin that has to be recorded. This occurred in March 1959 when the 700 cc Enfield engine was fitted to the Berkeley sports car to produce a vehicle with considerable get-up-and-go. The Berkeley firm built caravans and with the seasonal demands for these, looked to vehicles as means of keeping their workforce fully employed all the year round.

They began production in September 1956 with a very sleek sports car whose fibreglass body had faired-in headlights, a purposeful radiator grill and a soft top. The early cars used a 328 cc twin cylinder Excelsior two-stroke engine and gearbox unit to drive the front wheels, later models having a third cylinder squeezed between the two to increase the capacity to 500 cc. The gearbox had a reverse gear and the specification included 12 in. wheels, independant front suspension and hydraulic brakes.

Late in 1959 Berkeley modified the rear half of the car to a single wheel which gave them an instant 3 wheeler. Unlike many such of the period which were bubble-cars, underpowered and mainly disappeared with the advent of the Mini, the Berkeley had two wheels at the front as did all performance 3 wheelers and it went well.

The body was really too light and while this helped the performance, it also meant that the weight of driver and passenger could deform it enough to prevent the doors from closing. In addition to the normal soft top, a hard top was available as an alternative.

The fitting of the Enfield engine to the car pushed the performance up and two versions were offered. One used the Super Meteor with 7·25:1 compression ratio, single Amal and 40 bhp at 5500 rpm and this was the B95 model. The second had the more powerful Constellation unit with 8:1, Amal TT and 50 bhp at 6250 rpm and was the B105 model. In both cases a 12 volt electric system was fitted with a car starter mounted outboard of the engine sprocket and a belt driven dynamo fitted behind the cylinders. To accommodate the increased engine height, the bonnet line was raised, also that of the front wings. These had standard 7 in. Lucas headlamps fitted and to some extent the car lost its sleek look with the absence of the faired-in units.

In some ways the cars were crude but they provided transport, fun and very real performance with a top speed around the ton. Allied to this was the low down pull of the twin which made for an exciting vehicle at a time when few standard cars did more than carry people about.

Sadly the firm was not to survive due to a combination of circumstances rather than one single event. Thus their need for the Enfield twin went but they were not the only company to use it as the Frisky Sprint of 1959 also employed the Constellation engine to drive its 12 in. wheels. It had chain drive to its differential, a single long wishbone and coil spring damper at the front with a swinging axle at the rear located by a strut. It too failed to survive very long.

If they had they might have had an influence on the fortunes of the Enfield company but as it was they could only provide an interesting sidelight on the life of the big twin.

Constellation 700 in 1962, similar to but not identical with the Super Meteor. Siamezed exhausts and twin carburettors. Note ears on toolbox lids

6 Interceptor

The Interceptor was introduced onto the British market late in 1962 following a ready acceptance abroad. It had been built at first for the USA where the cry was for more engine capacity to provide effortless high speed cruising and even better acceleration. For a year or so, much like the 750 cc Norton Atlas, all production headed west for the American turnpikes but finally the immediate demand was satisfied and some models came into the home salesrooms.

The new machine was closely based on the Constellation but stretched out to dimensions of 71 × 93 mm which gave a capacity of 736 cc. The lengthened stroke called for a new crankshaft which was still a one-piece nodular iron casting, dynamically balanced and running on two substantial mains, one ball and one roller. Alloy rods with split shell bearings continued the Enfield tradition, as did the deeply spigoted iron barrels, but the new alloy heads carried two Monoblocs mounted parallel to one another. New was the seal between each head and barrel with the upper end of the bore chamfered to take a Cross triangular section steel ring. The push rod tunnels were sealed with steel pressings to which were bonded rubber rings, the assembly fitting into recesses machined in the underside of the head.

The 1963 Interceptor which was based on the Constellation and built to a full sports specification and finish

ROYAL ENFIELD

The lubrication system was still based on the twin pumps, as always, but they reverted to their earlier speed with a change back to a single start drive worm but with larger diameter pump plungers. This made their operation slower but surer and less prone to cavitation. The supply side was still the same with the oil leaving the tank via a combined drain plug and filter on its way to the pump. From this one line fed the crankshaft and the other the rocker spindles, and for these there was a change in the form of flexible feed pipes. For the scavenge effect both sides of the pump took from the sump and returned, with the overflow from the timing side fed from the exhaust camshaft tunnel, straight back into the tank.

The compression ratio remained at the 8:1 of the Constellation but the twin carburettors were parallel and flange mounted to short inlet stubs. The float chamber was still only fitted to the left unit which supplied the right via a flexible pipe. Siamezed exhaust pipes were fitted and the engine had two facilities already in use on the 700 cc motor. The first was a rev-counter drive taken off the exhaust camshaft sprocket nut which was specially made to couple with a right angle drive box bolted to the outside of the timing cover. All the covers were machined to allow this and, where the rev-counter option was not taken up, the hole was blanked off with a small plate. The second facility was that the left side camshaft bushes were mounted in circular housings which were held in the crankcase by three screws. Their removal allowed the camshaft to be withdrawn and changed without the need to part the crankcase halves.

There were three camshafts which became available, the standard one, a sports, and a super-sports designed for maximum speed and without quietening ramps. Alternative pistons giving a reduced compression ratio of 7·25:1 were also available and the figure could be further reduced, with either piston type, by the use of one or two compression plates. In the standard UK form power output was 52·5 bhp at 6000 rpm. Ignition continued to be by magneto but driven by a duplex chain, and fitted with manual control, while electric power was generated by an alternator.

The clutch was modified a little so that it contained three heavy gauge springs which were simply compressed by the pressure plate and three lighter ones between them which could be adjusted to ensure the clutch lifted true and clean. The gearbox was the normal four-speed type used by Enfield but had a light alloy bracket bolted to the rear of the shell which attached to the frame just below the fork pivot.

The frame and cycle parts were those of the Constellation but fitted with a conventional rear mudguard, in place of the rear enclosure, and a new prop stand without the Enfield spike. It was positioned much further forward and at last they

Above **A group of 1968 Interceptors fitted with 2 gallon tanks and cooling discs on the front hubs**

Left **The 736 cc engine of the 1963 Interceptor. The breathers, the camshaft bush housings and the toolbox lid ears are all clearly shown**

had a prop that was satisfactory. Wheels and brakes all came from the 700 with the tank but the seat did not and was a two-level type which failed to hug the rear mudguard as stylishly as before. The finish was in polychromatic blaze, a reddish tangerine, for the tank top and side covers, with chrome tank sides and mudguards, black frame, forks and centre box, and the usual Enfield tank badge. The UK price was £310 and the machine was available with the Airflow fairing.

Late in 1962 *Motor Cycle* had one out on test to record its ability to cruise at over 90 mph indefinitely. The top speed was 108 mph but this was of less importance than the sheer grunt of the engine which got the machine off the mark very smartly even on the 1 in 3 test slope. This ability to pull was to an extent highlighted by the gearbox for there was a big drop down to third so the pulling power was needed. On this test the brakes worked well and the handling was no problem, despite the weight and bulk of the model. The modified clutch also worked well and proved to be an improvement on the earlier type, while the gearchange was typical Enfield with short movements and a stiff lever. In order to position it so that it could be moved without lifting the right foot from its rest, some filing of pedal and footrest was needed. A more pronounced bend in the pedal would have given the same effect. The kickstart pedal was also moved round on its splines to increase the length of swing as cold starting proved to be inconsistent. When hot there was no problem. Once again it took great strength to lift the machine on to its centre stand but the new prop stand worked very well and was used as far as possible. All told the Interceptor carried on the traditions of the Meteor and Constellation that had preceded it.

The Interceptor continued in this form for

1963 but for the next year was available in a standard build as well as the normal one, which became the de-luxe. This had its electrics changed to 12-volt and the alternator output controlled by a zener diode mounted on a heat sink plate positioned under the combined battery and toolbox centre section. The magneto remained but with automatic advance and this was also fitted to the standard model which retained the 6-volt electrics. Its price was held down by the use of painted mudguards and tank panels, while both models were available with the Airflow fairing or to sidecar specification. The latter replaced the final Constellation and used the same low compression engine, the ratio being 7·25:1, sidecar tread tyres, lowered gearing, stiffer suspension and forks with less trail but fitted with a steering damper.

Motor Cycling carried out a double test in the middle of the year, first as a solo and then with a Watsonian Palma sidecar attached. In the first guise it again proved how suitable it was for high-speed cruising, although the hard front suspension and soft, under-dampened rear units were not the best combination for bumpy corners. The clutch again showed itself to be good and the gearbox the same with the usual short, heavy pedal movement. Maintenance proved to be as easy as was expected of an Enfield but the toolbox lids very hard to replace. The change to 12-volts brought with it a more powerful headlamp and much improved light beam.

Fitted with the sidecar and all the relevant machine alterations, the follow-up test showed the Interceptor equally at home on three wheels as it had been on two with an ability to accelerate quickly up to its top speed around 75 mph. With consumption at 45 mpg under hard driving conditions, it made an economical way of combining sporting performance with a means of carrying a family.

For 1965 the standard version was dropped and all Interceptors built to the USA pattern to assist production. Since their inception the US models had always had a frame with longer swinging fork, as this allowed the use of a fatter rear tyre and the extended wheelbase gave more stability at speed. At the same time the casquette was finally discarded and a separate headlamp fitted with the instruments and switches in a fascia panel above it. At the rear, Girling units were fitted in place of the usual Armstrongs, and the UK buyer had the option of the US style bars. As always, the Airflow was an option and the machine could be had in sidecar specification, although this no longer included the special fork ends to reduce the trail. After providing them for

Left above **Firing up a 1968 Interceptor, a task made easier by the adoption of coil ignition. Separate exhaust systems**

Right **1969 company advert for the Series II Interceptor showing the low mounted oil cooler fitted as standard for 1970**

We only supply the bike..
Since its introduction twelve months ago, the SERIES II INTERCEPTOR has achieved an impressive reputation. Built to meticulous standards, its rugged good looks, immaculate handling, enormous torque throughout the rev. range and train-like reliability have proved an intriguing combination for the man who doesn't just follow the crowd. It's been proved in battle too; numerous road race successes without a single mechanical failure have publicly confirmed the confidence we have in our product. For 1970, the INTERCEPTOR specification reads basically as before, but there's a significant difference —
FOUR BIG EXTRAS—and still with the 1969 Price Tag!
The oil cooler, carburettor, air cleaner, skid plate and pillion passenger hand rail, previously offered as separate and additional extras, have now become standard equipment, to further enhance the individual character and powerful personality of this thrilling mount.
ROYAL ENFIELD
Interceptor
Further information from :
ENFIELD PRECISION ENGINEERS LIMITED
UPPER WESTWOOD, BRADFORD-ON-AVON, WILTS.
Tel.: BRADFORD-ON-AVON 2166

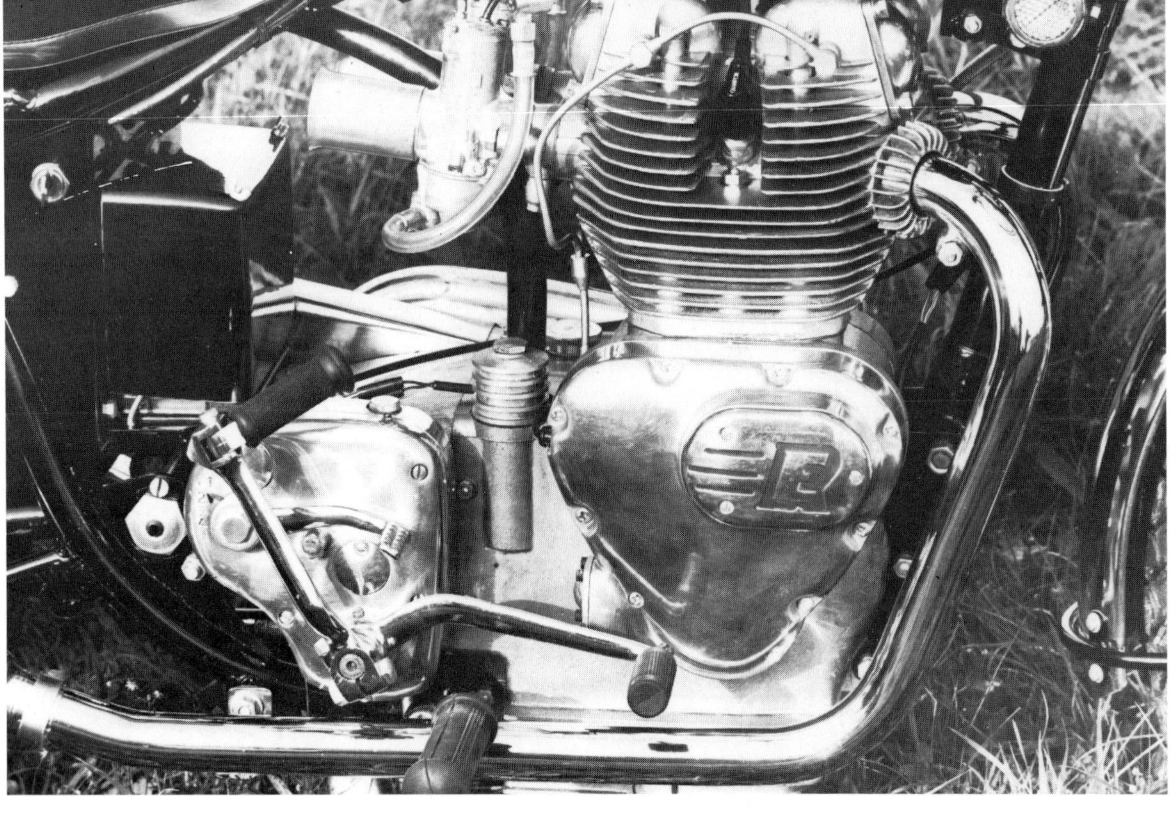

Left **The home market Series II for 1969, a handsome machine in the Enfield twin tradition**

Left below **The wet sump engine of the Series II with points in timing case and single oil pump to its rear**

Below **Left side of the Series II Interceptor fitted with twin parallel Concentrics. In the abbreviated stark style of the period**

so many years the firm said they were not needed.

From early in 1966 the Interceptor was removed from the home market as the factory concentrated on the 250 cc models, and what big twins they did produce went to meet export contracts in America.

During the early and middle 1960s the company had gone through an involved history of moves and purchases. Since 1940 an associated firm, the Enfield Precision Company, had been engaged on the production of secret weapons and high precision parts for war equipment. This work was carried out at Bradford-on-Avon in Wiltshire and the factory was set in underground caves in a disused

The Rickman Enfield in 1971 clearly showing the footrest support welded to the exhaust pipe

quarry, these extensive caves having been used for storing art treasures during the war. In 1958 Enfield bought the factory from the Ministry of Defence and in 1962 the Enfield Cycle Company was itself purchased by E & H. P. Smith, who also owned Alpha Bearings.

During 1966 the company found itself in a steadily worsening financial position and the range of machines began to contract with the last of the small singles being withdrawn in January 1967. Then, in March, came the news that the then recently formed Norton Villiers company, whose parent was Manganese Bronze, had taken over the Enfield Cycle Company from E & H. P. Smith with all rights being acquired for £82,500. Neither the Redditch or Bradford plants were included in this deal, which also gave Enfield Precision Engineers a contract to build Interceptors for export to the USA.

This gave the Bradford plant some work while Redditch became the home of a new company called Enfield Industrial Engines with Tony Wilson-Jones as chairman and Jack Booker as managing director. This did not last too long and eventually the famous Enfield Redditch works were sold to the local Development Corporation.

The take-over gave Norton Villiers about a 30 per cent share in Enfield-India and initially distribution rights for the Interceptor. These were, however, soon returned to the Enfield concern along with the spares and service business. Further complication arose with the Velocette company taking over the remainder of the spares and service work from their Hall Green factory.

Ultimately the manufacturing rights for Enfield and Velocette were acquired by Matt Holder who had held those of the Scott for many years and also those for the Vincent. As managing director of Aerco Jigs and Tools he was to continue his interest in motorcycling by building Scotts and making spares for Velocettes up to the time of his death in 1981.

Meanwhile, back at the factory, Interceptors continued to be built in small numbers and in late 1967 reappeared on the home market. They were much as before but had changed to coil ignition, Concentric Amals, separate exhaust systems, and were fitted with a 2-gallon tank and cooling discs on the front hub, these having large pierced holes in them.

1970 Rickman Metisse fitted with the Series II engine. Typical immaculate frame finish which is standard for the Rickman brothers

Late in 1968 came the Series II model and this was considerably revamped with only the frame and rear hub escaping unchanged. The engine was based on the older unit but the first fundamental difference was that a true wet sump lubrication system was fitted with a single pump. Thus the oil lived in new ribbed castings which carried it under the crankcase rather than behind it, and a completely new timing cover carried the rear half of the old pump drive shaft. The familiar Enfield pump fitted into the rear of the cover, rather tucked out of sight, and the oil supply was directed at 60 psi to the big ends and 15 psi to the valve rockers. It all drained back to the sump which contained a fine mesh gauze filter to protect the pump, while the traditional felt filter and its magnet were moved from the timing cover to a vertical housing behind the cylinders. The filler cap continued to carry a dipstick and the capacity was slightly increased. For really high speed work or for racing, an oil cooler was available as an option and this mounted in front of the crankcase, rather low down and in line of any debris thrown up by the front wheel.

The new timing cover was of triangular shape and no longer extended back for a magneto drive. Lucas capacitor ignition was fitted and triggered by contacts opened by a cam on the end of the exhaust camshaft. The contact points were housed in the timing cover with a small alloy casting over them, while the left end of the exhaust camshaft drove the rev-counter.

Twin Concentric carburettors were fitted with the option of a massive air cleaner available and the machine had separate exhaust systems. The primary transmission was as usual with a duplex chain but the clutch was very much revised and proved to be totally able to cope with the power and the torque. The gearbox was the usual Albion still complete with neutral finder and curved gear pedal.

At the front the old frame carried Norton gaitered forks and their single leading shoe front brake, which immediately allowed any of the special twin leading shoe options to be fitted. Its 8 in. diameter was to provide the leverage that the Enfield units had lacked. The rear hub complete with brake, cush hub and qd facility remained unaltered.

On to the frame went a 2 or 4-gallon tank, a

1971 Rickman Enfield with Interceptor engine—a big lusty motorcycle for men. Disc brakes were fitted to both wheels

new seat and revised details such as mudguards and silencers. After years of casquettes the instruments were just bolted to the top of the forks and the ignition switch to a frame panel on the right below the seat. If required, a skid plate could be fixed under the engine and a rail to the back of the seat.

And it continued to perform as always. With no need to make the engine turn over rapidly and with a crankshaft of weight and in dynamic balance, there was little trouble with vibration. The gearbox was not used or needed to any great extent and the rest of the machine matched the style of the engine. The new forks worked suprisingly well with the old frame and the only snag was the forward position of the footrests which inhibited high speed cruising.

For 1970 the options of air cleaner, oil cooler, skid plate and seat rail were all fitted as standard without any price increase, but the days of the company were numbered. The whole of the British motorcycle industry had fallen on to hard times and, as others were to be forced to do in later years, Enfield finally had to stop motorcycle production in the middle of the year.

Although this meant the end of the Interceptor as such it was not to be the final end of the line for the Series II engine. During 1969 the Rickman brothers became involved with the Enfield marque and the first result was a Rickman Metisse fitted with the Interceptor engine and first seen at the Sporting Show held in Victoria, London early in 1970. This combined the tough Enfield engine with the immaculate Rickman chassis to produce a machine with even more punch, thanks to the reduced weight. The frame was of all-welded construction with duplex tubes throughout and finished by nickel plating. The swinging fork was mounted on an eccentric pivot to provide chain tension adjustment and the front forks were Rickman. Both hubs carried disc brakes, which at that time was unusual on the front and very rare for the back. They were cast iron with Lockheed hydraulics and the wheels were spoked with alloy rims. The seat had a small step in it with both halves being flat and it sat on a fibreglass moulding which extended back into a tail unit and rear mudguard. The tank was also fibreglass with quick action filler cap and a shape that was not to everyone's liking. The two coils nestled just behind the gearbox with the area above them boxed in for battery and tools.

The most talked about feature was the mounting of the footrests, each of which was clamped into a small vertically mounted tube which was welded to the top of the exhaust pipe. Odd, but it seemed to work.

The performance was little changed but the handling was very good indeed and was not faulted by *Motor Cycle News* when they rode the prototype in April. Their test highlighted the need to tilt the ends of the exhausts up to avoid them dragging on the ground. Once again the smoothness of the big vertical twin was emphasised which showed that the balancing work did pay off.

The next spin of the wheel brought in an import/export firm, P. Mitchell of Birmingham, who acted for Floyd Clymer of the USA to buy about 600 1969 Series II engines and have them built up with Italian frames. The complete package was sold in the USA as the Indian Interceptor and thus revived the link of the 1950s between the two company names.

This deal died with Clymer in January 1970 leaving Mitchells with many engines so they commissioned Rickman to build a few machines in their usual style to show the trade. These were then used to set up an exclusive deal with Elite Motors at Tooting in London and Rickmans went on to build a further 130 machines of which 26 were shipped direct to Canada and the rest sold by Elites as the Rickman Enfield.

The personal export price was £625 and so on a par with other machines of the times which could not boast the disc brakes or the superb nickel finish. The machines all sold, many of them to the USA, and most owners liked them for the rugged power, beefy acceleration, handling and brakes. Less popular was the hard suspension and rather forward footrests which, with wide bars, made high speed cruising a burden.

The Rickman Enfield was one of the last of a particular breed of motorcycle. By 1972 the whole scene was changing. The CB750 Honda, 500T Suzuki and H2 Kawasaki had appeared, while the Z1 was imminent. Racing was being dominated by the TD and TR Yamahas soon to be superseded and out-paced by the TZ. Motorcycling for the public had spent 10 years in the doldrums but was poised to take off with sales that were to increase in leaps and bounds as it became respectable in the 1970s. Into this new era of refined fours with electric start and every convenience the lusty Enfield was something of a dinosaur. Not really a relic from the past, for it could still ease its way past the 110 mark, but a machine from a different and simpler mould.

It was a machine for a man's world and no-one denied it. In no way could a slightly built woman hack 736 cc of Interceptor into life, however enthusiastic she might be. But the men mourned the end of the Enfield line and what it stood for.

Some hope of a revival occurred at the 1979 Earls Court Show when Jim Norris, an Enfield enthusiast with a motorcycle business in Wales, showed an Interceptor built up from spares in its Series II form. He had hoped to go into very limited production but this was not to be and no more was heard of the venture.

So ended the Enfield twin line which had begun with a tourer and finished with a real big motorcycle.

The rear disc brake of the Rickman Enfield with Lockheed caliper. Very rare in 1971

7 Clipper, Crusader, Turbo

One of the factors that enabled Enfield to produce a good range of machines at very competitive prices was their ability to combine a small number of major assemblies in a variety of ways to suit differing needs. Thus frames, forks, wheels and cycle fittings were used over and over again with subtle variations to give different effects, while engines and gearboxes shared many common parts all of which greatly reduced the spares stock holding at the factory and the dealer. At times it also led to some odd machines being built up either by the factory to reduce stock levels, or for customers at home or abroad who had special needs. Thus it is not unusual to come across Enfields that have non-standard features, although factory built, and special colours are also to be found. If anyone came knocking on their door with a specification, Enfield would always manage to cobble something together to get the order.

Late in 1953 another throw of the dice produced a new Royal Enfield but one that looked familiar and common to the range. At the front it looked Bullet or Twin, amidships it was model G, but behind the gearbox became Bullet again. It was a machine built to fill the gap that had existed since the war between the 125 cc two-stroke lightweight and the much heavier 350s in either G or Bullet forms. It was a size that Enfield had not built since 1939, 250 cc, and its appearance was part of a gradual trend back to that capacity. In 1953 there were few machines of that size to fill the gap between models powered by the ubiquitous 197 cc Villiers engine and the many nifty 350s of the times. During the next decade 250 cc was to become a very significant motorcycle capacity, partly due to it being the limit for learners for many years and partly as the performance level reached a point where it became interesting and even exciting instead of drab.

The rigid frame model S of 1954, in effect a smaller version of the G with separate gearbox. Seldom even mentioned.

The new Enfield was called the 'Clipper' and was aimed at riders who wanted a simple, economical single with modern specification and reasonable performance. It was based on a separate engine and gearbox layout but with swinging fork rear suspension for the frame.

The engine was a smaller version of the G and J units down to fine details. It was based on dimensions of 64 × 77 mm so was of 248 cc with a compression ratio of 6·5:1. Construction followed that of the larger models with cast iron head and barrel, alloy crankcase with integral oil tank, filler cap at the front of the crankcase, oil

Better known was the 250 cc Clipper with sprung rear wheel. The separate gearbox was retained on this model

feed up to the rocker box and single alloy cover on top of the engine. From the right the timing gear, oil pump and filter looked exactly Enfield.

From the left the appearance was not quite the same for, unlike the larger models, the 250 was fitted with an alternator and coil ignition. The Miller alternator was bolted to the end of the crankshaft and enclosed in the primary chain-case which was enlarged at the front to accommodate it. The ignition points were fitted very neatly in a housing mounted on the magneto platform and strapped down in the usual way. This made them very accessible under a simple cover while being well out of harm's way. The rectifier of 1953 was not the compact device it was to become in time but a massive assembly of four circular plates which was hung just below the front of the tank where it benefitted from the cooling wind. The electrical system was controlled by a combined ignition and lights switch mounted on the top of the casquette on the right, where it was balanced by a Miller ammeter with integral warning light.

The transmission of the Clipper was standard Enfield with chain drive to a four-speed Albion gearbox mounted separetly from the engine in plates. It was, however, a lighter version and differed from the others in that the clutch cable ran round behind it and no neutral selector was fitted. It retained the high pedal pivot of the larger machines but this was further forward, nearly directly over the footrest, so easier to use.

The frame was a mixture of model G and Bullet with the front down-tube just reaching to the top corner of the crankcase. Its open diamond form continued with the top-tube splitting into two loops which swept back to support the tops of the spring units and down under the gearbox. From them pressings supported the swinging fork pivot and the gearbox plates which were also tied to a saddle tube. At the front went Enfield telescopic forks complete with the casquette, as mentioned, although this was a smaller version than that used on the other models and did not feature the twin parking lights.

The detail cycle parts followed convention with an unsprung front mudguard and the very easily detached Bullet type at the rear. Toolboxes were fitted on each side in the rear frame loop with the air cleaner box on the right and battery on the left just in front of them. The carburettor was Amal and the exhaust system ran low down on the right to a tubular silencer. A saddle was fitted as standard, although a dualseat was available as an option with pillion footrests which attached to the rear fork pivot supports.

Also made available was the same engine and gearbox fitted into the model G cycle parts to produce a cheap, rigid frame version called the model S. This retained the Miller generator and coil ignition from the Clipper and the separate headlamp and rear stand of the G. It made a useful utilitarian package but was not kept in the range for long and left it late in 1954.

The 1955 250 cc Clipper with oval air cleaner and massive rectifier under the front of the tank. Casquette but no pilots

The Clipper sailed into service without any problems so that for 1955 the only changes were those common to other models. Thus the gearbox end cover was cleaned up to enclose the clutch lever and bring the gear pedal pivot down to be concentric with the kickstarter. In addition, the internal gearbox ratios were altered to make them wider apart and the air cleaner box changed to the oval shape. The price remained unaltered at £162.

It was in this form that *Motor Cycling* road tested a Clipper late in 1954. It proved to be good for 62 mph with acceptable acceleration and an ability to cruise at 55–60 mph for long periods. Such a speed was quite adequate for many journeys in pre-motorway times and over the test the magazine recorded 95 mpg. Handling, comfort, and steering were all praised for they matched those of the larger machines, while the brakes worked well. For a utility mount the Clipper was well founded and this made for a very pleasant machine that rode over the bumps in the road at a time when all too many hopped from crest to crest. The lack of a lifting handle or a prop-stand was a shortfall, as was the slight fork judder experienced under heavy braking. Otherwise, the only complaint concerned the location of the tyre pump on a mudguard stay where it suffered from road dirt, while being obvious to the light-fingered.

The 250 cc Clipper proved to be sufficiently successful to need few changes for 1956. Those made were to the ancillaries with the fitting of an Amal Monobloc and a change to Lucas electrics. This entailed the fitting of a new alternator, the movement of the ignition switch to the front of the right toolbox and a new contact breaker assembly which plugged into the housing on the magneto platform.

With the 250 came a big brother of 350 cc using the same name and principal. It used the model G engine dimensions of 70 × 90 mm and the same format of separate engine and gearbox in swinging fork frame. It was equipped and looked like the 250 but differed in respect of a few details. It used a Lucas mag-dyno and, due to the space restriction, an old type Amal was fitted. Compared to the 250 the tyres were a little fatter but smaller in diameter, the casquette had the twin pilot lights and the gearbox the high pedal pivot and neutral finder. It was a few pounds more than the 250 but £21 less than the Bullet, for those who wanted an economy machine. As such it continued to have a saddle fitted, although a dualseat was available as an option, as was a prop-stand.

The Clippers were fine for the utility rider but even the ardent Enfield fan had to admit that

Right **The company exhibit in the 1956 Redditch Carnival where it took first place in the local manufacturing class**

Below **The unit construction Crusader engine used by Royal Enfield for many years. Camshaft on left with oil pump, points on right**

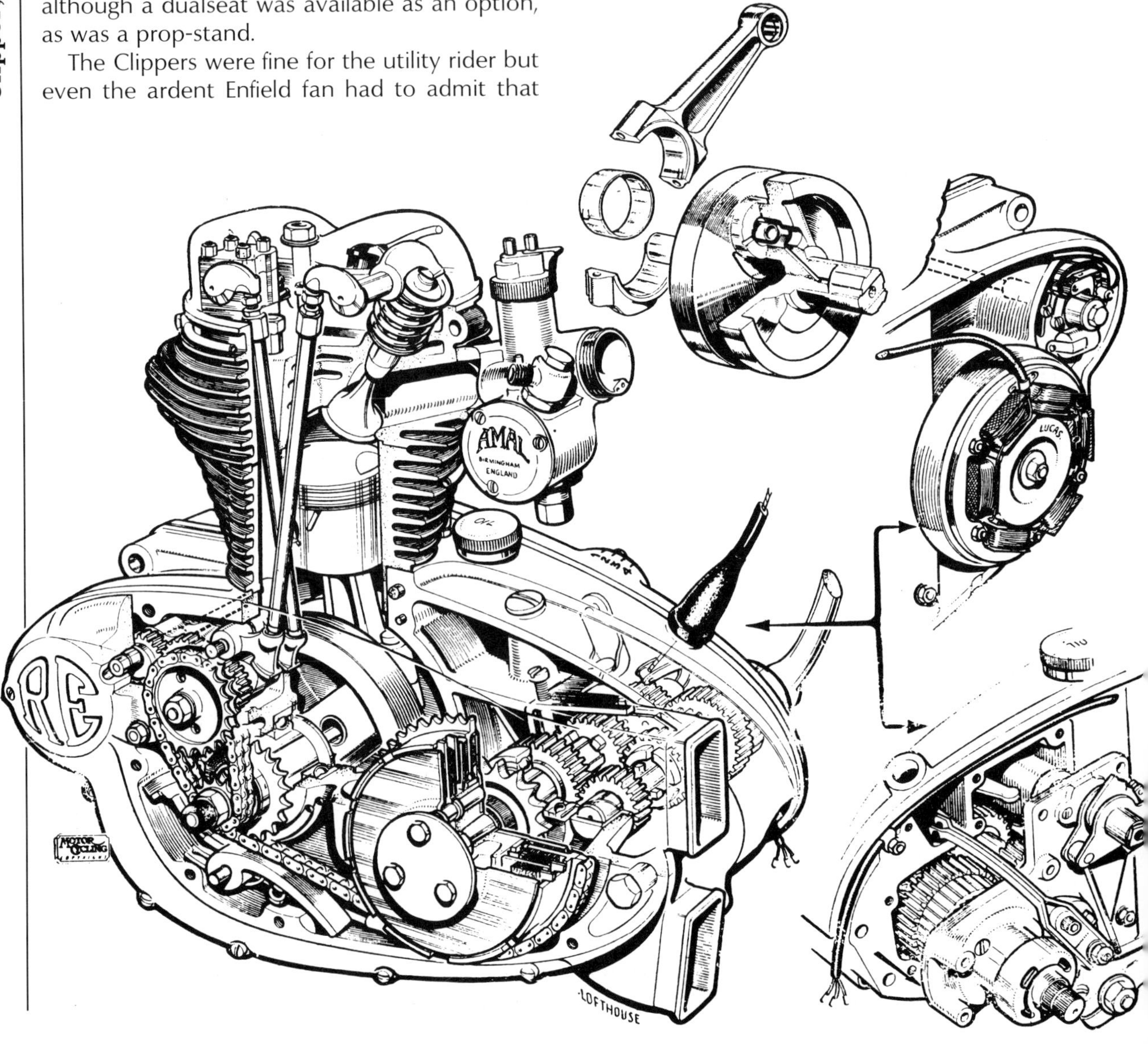

their appearance was hardly attractive compared to the Bullets, due to the odd combination of parts used in their make up. In August 1956 came news of a completely new Enfield to change that and provide 250 cc riders with an up-to-date specification in a sleekly styled package.

The new model was the Crusader and of 250 cc. The engine and gearbox were built in unit, it had small wheels, an enclosed rear chain, and all told offered some very nice features for a reasonable price. Although the engine of the Crusader was built in unit with the gearbox and they shared the same crankcase castings, the two assemblies lived in separate chambers. A third chamber surrounding the gearbox to the rear of the crankshaft housed in the engine and primary transmission oil in normal Enfield way, although the quantity was down a little.

The engine itself was based on dimensions of 70 × 64·5 mm and was only mildly tuned with a compression ratio of 7·3:1 and cast iron head and barrel. The crankcase was split vertically along the cylinder centre line and this item was deeply spigoted into it. It carried the one-piece cast iron crankshaft in two massive main bearings, that on the left drive side a ball race and that on the other a roller. The crankshaft located on the ball race and comprised two substantial flywheels with short fat mainshafts. These overlapped the massive crankpin diameter to an appreciable extent to provide a very rigid and stiff crankshaft.

The rod was in light alloy and the split big end was fitted with shell bearings. Fitted bolts were pegged into the rods to prevent rotation and aligned the cap before the two nuts were torqued-up and split-pinned. The piston was conventional with a small dome and the usual scraper and two compression rings. The gudgeon pin ran directly in the the rod, was taper bored and retained by circlips.

The piston ran in an iron barrel and this was spigoted up into the head. The push rod tunnel was cast-in and a neat touch was its restriction at its lower end to two holes positioned to guide the push rods on to the cam followers during assembly. The head was also cast iron with pressed-in valve guides and split rocker housings following normal Enfield practice. Duplex valve springs retained by conventional collars and collets were fitted and the whole valve assembly enclosed by an alloy cover held by a single nut. Head and barrel were held by five studs screwed into the crankcase with a paper gasket under the barrel and a copper-asbestos one between it and the head.

The valve gear was conventional in that a half speed camshaft ran in bushes above the crankshaft and was driven from it by a duplex chain. The cams lifted bellcranks and each of these worked a push rod of steel tubing with a ball end pressed into the bottom and an adjustor at the top. To prevent the end of the tube from belling out, it was chamfered to fit into an undercut shoulder on the end. In turn the adjustors at the tops of the push rods made contact with the rockers, which were stampings, and these opened the valves each of which was fitted with a hardened cap.

What was unconventional about the valve gear was that it was on the left of the engine, an unheard of position on an English machine on which the timing gear had been on the right for decades. The camshaft was directly over the

ROYAL ENFIELD
RE

ROYAL ENFIELD
RE

Left **The 250 cc Crusader launched in 1956. This example is fitted with the optional panniers and legshields. The chain enclosure is standard**

Left below **The new Crusader from the right. Single engine head steady runs back to the frame, rectifier is under toolbox**

Right **For America the Clipper became the Indian Hounds Arrow in 1955. Hardly the sump for off-road work**

Below **The 350 cc Clipper of 1957 with separate engine and gearbox of model G in springer frame**

Above **The 1958 Crusader with two-tone finish for fuel tank, mudguards and toolbox lids**

Right **Crusader with Airflow fairing fitted, a very real attempt to offer good protection to the rider. The matching front mudguard runs outside the fork legs for maximum effect**

crankshaft and the duplex drive chain lay outboard of the single strand primary. Chain tension was set by an adjustable slipper on the rear slack run. The method of driving the crankshaft sprocket was ingenious and avoided the need for a slender mainshaft, despite the small size of the sprocket which was little more than the substantial mainshaft. This shaft was drilled and tapped axially and also machined across its end with an offset recess. A stud was screwed into the shaft and carried the sprocket which was machined with a tongue to match the shaft. A nut on the stud held it in place and also served to retain the engine sprocket on its spline, clamped up against the ball bearing main to locate the crankshaft. The stud was also hollow and extended into a seal in the chaincase cover to a recess supplied with oil to feed into the big end.

The crankshaft was drilled to feed this oil across the crank web into the pin, which was hollow. Into the crankpin bore was pressed an alloy tube with reduced centre section to form an annular space into which the oil was supplied and from which it exited to the bearing surface via two holes.

The camshaft was also involved with the lubrication system and the ignition. Pressed on to the shaft was a small gear and this drove forward via an intermediate with hunting tooth to a shaft that ran across the front of the engine. A cover plate was fitted over the outer ends of the three shafts and at its front carried the two oil pumps. These were based on the Enfield system of a plunger working in a cylinder in an oscillating block and in this case the two plungers for delivery and scavenge lay side-by-side and both were worked by a common eccentric pin pressed into the end of the forward cross-shaft. A single block was held into the housing by a spring and suitable holes formed the inlet and outlet ports as the plungers moved with it. The scavenge plunger was larger than the delivery.

Oil was drawn from the tank formed in the crankcase and passed through a felt filter fitted into the top of the left crankcase half just behind the barrel. From there it went into the crankshaft via the hollow stud in the mainshaft. On the scavenge side oil was collected from the crankcase sump and returned to the tank. A spring-loaded ball valve in this return diverted some oil to an external oil pipe which attached to the crankcase to the right and rear of the barrel and connected to the cylinder head at two points with banjos. These supplied the rocker spindle bearings and the overflow drained down to the timing chest. In this a weir level ensured that cams, followers, and gear train were adequately lubricated and the overflow fed into the primary chaincase. Again, a weir was used to ensure enough oil for the chain and, in this case, the overflow was directed into the sump for collection. The oil tank compartment was vented into the chaincase, which in turn was connected to the crankcase. Breathing was via an outlet union fitted on to the top of the crankcase in front of the barrel with an exit tube trapped to form a baffle to keep oil in while letting air out.

The cross-shaft also drove the cam for the coil ignition system and had a mechanical advance mechanism built into the assembly. The contact points plate was attached to the side of the right crankcase and access for adjustment gained via a small cover. The right end of the crankshaft carried an alternator rotor with the stator bolted to the crankcase outer wall.

The primary drive was by single strand, slipper tensioned, chain to a three-plate clutch lifted by a lever pulled by a cable exiting through the top of the right crankcase via a rubber sealing grommet. The gearbox was an Albion laid on its face so the layshaft sat behind the mainshaft. In other respects it was of standard form, although the gear pedal was linked internally to the positive stop mechanism and gear indicator. The left gearbox bearings were fitted into the wall of the crankcase on that side but on the right were carried in an outer plate which bolted to the outside of the right crankcase half. This plate carried the clutch lever and gearbox selector mechanism and it, together with the alternator and points, were enclosed by a single outer cover with access plate for the points. This matched well with the primary chain cover on the left, both being held by a series of screws.

The rear chain was also totally enclosed, with ports for the two chain runs being cast into the rear of the left crankcase half. These connected to the chain ducts with flexible bellows-type plastic gaiters which clamped to both parts. The rear sprocket was accessible on removal of a cover over its rear half and this attached to the chain ducts. For assembly purposes a small hole was provided above the gearbox sprocket to assist in guiding the rear chain round it, and this was sealed by a plate.

The engine breathed in through an Amal

Above **The 1961 Clipper of 250 cc based on the standard Crusader. A complete contrast to the earlier model with separate gearbox**

Below **A Crusader Sports from 1959. It has lower handlebars and footrests further back than the standard model**

Above **The standard 1961 Crusader with very deeply valanced front mudguard and new style casquette on which the pilots are further forward**

Below **The 350 cc Clipper of 1958 with semi-unit construction Bullet type engine in Crusader form of cycle parts. Note saddle**

Monobloc and out into a downswept exhaust pipe and low level tubular silencer. The exhaust port was offset to the right to suit this location but the inlet was on the engine centre line and led straight into the cylinder.

The power unit was fitted into an open type frame with single top and down tubes, the latter bolting to the front of the engine via plates. From the rear of the top tube two smaller tubes ran back and out to support the tops of the rear suspension units. They then turned down to run slightly forward to pass behind the rear fork pivot and then curl under the gearbox. They were bolted to the engine unit at this point and also from forward extensions of the fork pivot support plates.

The rear swinging fork was completely conventional, as were the lightweight telescopics fitted at the front. The single rate springs were anchored at top and bottom so could work in tension as well as compression. The forks carried the usual Enfield casquette fitted with a Lucas headlamp and the twin pilots. Also housed in the top of the assembly was a Smiths speedometer, the lights switch, and an ammeter.

Full width hubs, incorporating 6 in. diameter single leading shoe drum brakes, were fitted to both wheels, the rear one being quickly detachable while retaining the cush drive. Both were 17 in. diameter and shod with Dunlop tyres, ribbed at the front and studded at the rear. Each was protected by a substantial mudguard and that at the front was unsprung with a single rear stay. The rear mudguard with its boxed-in rear number plate was supported by a loop stay which ran along each side.

This combination of small wheels and large mudguards gave the model a chubby, solid appearance and this was further enhanced by the fuel tank and centre section. The first of these items was made from two pressings with a centre join and held 3 gallons. It was secured by a front bolt and rear clip on to a rubber bush in a manner used on other models. The centre section combined several functions into a box assembly with two side lids. Into it were located a Vokes air filter which attached to the carburettor with rubber trunking, an ignition switch on the left, a battery, and finally the toolkit. To its underside was bolted the rectifier.

For the rest, the electrics were standard English alternator with emergency start, the horn hung down just below the front of the tank, the machine came fitted with a dualseat and pillion rests, and the centre stand was at, or about, the point of balance and easy to use.

It was launched on to the home market at £198 8*s* 0*d*, with legshields available for a further £3 3*s* 0*d*, and formed part of the 1957 model line-up. The more utility 250 Clipper continued alongside it, together with the 350 cc version, and neither model had any changes. The Crusader was listed in a range of colours and in black.

At the end of the year the Crusader received some minor changes to the gearchange mechanism and the lubrication system, together with a Burgess silencer. Its finish became black as standard with the no-cost option of a two-tone finish which kept the frame, forks, centre section body and chaincase black, while allowing the tank, mudguards and toolbox lids to be in burgundy, grey or blue. In this form it went forward into 1958 but without the 250 cc Clipper which was dropped, while the 350 cc version received a major facelift.

The revised 350 Clipper was based on a Bullet bottom half fitted with an iron head and barrel. To this was bolted the usual four-speed Albion gearbox and this was driven by a duplex primary

Right above **1962 Super 5 with leading link forks and five-speed gearbox in place of the normal Crusader parts. Slight sweep back to exhaust**

Right **The 1961 350 cc Clipper, in effect a Bullet engine unit in a Crusader frame. Much confusion exists over types and names**

ROYAL ENFIELD
RE
144 BAB

ROYAL ENFIELD
RE 350

chain in an alloy case with its outer held by a single fixing. All stock Bullet features in fact. The engine had an alternator fitted to the left side and its ignition was by coil with the points in a housing in the old magneto position. This gave enough room for a Monobloc to be fitted, while the exhaust pipe carried the old type of silencer, not the new Burgess.

The frame was the usual open-loop type as used on much of the range and was fitted with modified Crusader forks and front wheel, the double lidded centre section and a saddle. As it was aimed at the utility market it was natural to use a combination of parts from other machines and, due to this, it had the casquette, and a dualseat was available as an option. However, at the rear it was fitted with a non-qd, non-full width rear hub. It was finished in black with chrome plated wheel rims and the tank was embellished with winged tank badges, also finished in chrome. It sold at £199 18*s* 3*d* so had a £30 margin on the more sporting Bullet, and over a tenner on the smaller Crusader.

Early in 1958 *Motor Cycling* published the first official road test of the Crusader and reported that it met its criteria of reliable and economical transport. It proved to be able to work up to 70 mph and record over 90 mpg on general runs, with 100 mpg within easy reach. This was helped by an increase in the compression ratio which had risen a little to 8:1. The Crusader started first kick and ticked over slowly and evenly. It had the usual light Enfield clutch and gearchange, handled well, was comfortable, and stopped when required to. With the weight it had to pull along it was not to be Enfield's most exciting machine but for day-in, day-out, use with easy maintenance it was hard to better.

The model tested was fitted with the optional legshields which, together with the deeply valanced mudguards, served to keep the worst of the wet weather of the test at bay. There was,

The 1962 Trials model based on Crusader parts and works experience. Well tucked in exhaust pipe and silencer, coil ignition

The 350 cc Clipper for 1962, alloy head, no pillion footrests and revised centre stand

however, more protection on its way for that month the Airflow fairing was announced and made its first appearance on the Crusader.

The Airflow was a very real attempt to give the rider weather protection and some carrying space in a rigid rattle-free unit that did not inhibit maintenance, enlarge the turning circle or create backdraughts. It was made in fibreglass in two main sections bonded together. The larger of these formed the main front nose and legshields, while the second smaller part formed an instrument fascia and cubby-hole bulkhead into which the main nose support fixing was bonded. This bolted to an extension tube which ran forward from the headstock and was supplemented by lower fixings below the cooling air duct in the moulding. These fixings were done with half tube sections bonded into the fairing which picked up on extended supports bolted to the front engine plates. The two parts were clamped together with a Jubilee clip on each side.

To enable the Airflow to be fitted the normal casquette was replaced with a small alloy fork crown in which the speedometer recess was blanked off by a round RE motif, actually a tank badge from the 125 cc two-stroke model. The speedometer was transferred to the fairing bulkhead and flanked by an ammeter on the left and light switch to match on the right. The Airflow also carried the headlight unit and this was held in a car-type mounting shell complete with three point adjustment for setting the aim of the beam.

The area behind the headlight formed the stowage compartment and access was by holes on either side of the forks. A curved Perspex screen was sealed and bolted to the front edge of the moulding and a styling line was taken from its centre to surround the headlamp and then run

Below **Crusader Sports for 1962 with one-piece seat mounting and rear mudguard**

Far right **The standard 1962 Crusader with deeply valanced mudguards. Handlebars and footrests also differ from Sports model**

back on each side to the edge of the legshield area and then down each side.

With the Airflow fairing came a really enormous sprung front mudguard styled to match and also made in two sections in fibreglass. The guard extended out to encompass the fork legs and was cross-braced with a bonded-in tube to ensure rigidity in the area under the fork crown to which it was bolted.

The two items when fitted pushed the machine weight up by only 18 lb despite their own weight of 33 lb, this being achieved by the removal of superseded items. The cost of the Crusader with Airflow was £257 1*s* 6*d*.

A few days later *The Motor Cycle* published a road test of one ridden over 600 miles in wintry conditions and reported that it gave exceptional weather protection. There were no real problems under gusty wind conditions and the steering was as normal except at a walking pace when it was a little heavier than usual.

Performance was very much as with the standard model but there were some complaints. These concerned the rather noisy exhaust note which would have been accentuated by the fairing, variable tickover speed, and heavy clutch. These last two may well have been due to the re-routing of the control cables with the fairing in place as this problem had cropped up on other models fitted with the same top crown with its cable guides. The suspension was considered rather hard but the low riding position and convenient controls were appreciated. The only point regarding the latter was a suggestion that the light switch and ammeter be swapped over to enable the lights to be more easily controlled by the left hand.

In March 1958 came yet another version of the Crusader, the Series 2 Clipper, which was an economy version built to get the price down to £189 19*s* 7*d*. This was done by omitting the front mudguard valances and full rear chaincase, fitting a chainguard in its place, and a saddle, a non-full width rear hub and by leaving the engine and gearbox castings unpolished. It was finished in black with chrome plated rims and gold lining on the tank and, at the price, provided a good replacement for the first Clipper model.

In the middle of the year Airflow production had reached a level where the firm was able to offer it on other models. The ones chosen were the new 250 cc Clipper and the Meteor Minor and in all cases it was available as original equipment or could be fitted to a customer's machine if this was returned to the works.

October brought the news of the changes for 1959 and yet another 250 cc single, the Sports Crusader. This was based on the standard model but with a new light alloy cylinder head giving a compression ratio of 8·5:1, a larger Monobloc, bigger inlet valve and hotter cams. With these went a new silencer and a 7 in. front brake to match the improved performance. The looks of the model were enhanced by a deep fuel tank with more capacity, footrests which were higher and further back than on the standard model, and sports handlebars which were swept down at their ends.

The three machines already in the programme had few changes, although the standard Crusader was also fitted with the light alloy head and new silencer. The compression ratio of the 250 cc Clipper was slightly reduced and both Clipper models were fitted with a dualseat as standard and finished in cherry red. For the 350, rear chain enclosure became an option and all

models could be fitted with the Airflow if required.

In May *Motor Cycling* road tested a Crusader Sports and acclaimed it the fastest standard roadster of its size yet tested by them, with a top speed of 78 mph. They also achieved 92 mpg over their entire test period which was an improvement over the standard model. In most aspects the Sports machine behaved much as expected, but with the benefit of enhanced power and the bigger front brake. This was further assisted by the low bars which encouraged a tucked-in riding style without making the rider's life uncomfortable. Their only snag was a reduced steering lock which made the turning circle rather large. Overall the picture was of a machine with a snappy performance and styling that was up to the minute for the time.

For 1960 an improved clutch was fitted to all the 250s, while the 350 Clipper was modified in line with the Bullet and, in fact, became a cheaper version of that model. This was done by fitting a light alloy cylinder head while retaining the same compression ratio, 17 in. wheels, raised gearing to compensate for this, a qd facility for the rear wheel, and a larger capacity fuel tank. The front brake remained as 6 in. but the rear was larger at 7 in., and the electrics were by alternator and coil ignition with emergency start. The revised 350 Clipper and both Crusaders were fitted with the new form of casquette common

Left **Cockpit view of Continental in 1963. Matched instruments fit neatly behind the flyscreen**

Below **250 cc Continental in 1963 with Italian-style petrol tank, cranked exhaust and half exposed rear springs**

to most of the range. The Airflow fairings were by then supplied by Bristol Aircraft who made them to the Enfield design.

Inside the Crusader series of engines a minor change had taken place to the rocker assembly. The rockers were changed to a two part construction with both carried in a single body secured to the head by two small studs at each end and a larger one in the middle. This modification lengthened the rocker bearings and improved the geometry of the arms of the rocker.

1961 brought standardisation of fuel tanks with that of the Crusader Sports being fitted to the other 250s. The Crusaders also received a minor alteration to the casquette, a new silencer, and the standard model was fitted with the same exhaust pipe bend as the Sports. The 250 Clipper had the casquette change along with its bigger tank, but the 350 version went forward unchanged.

The Motor Cycle had also had their chance with the Sports Crusader the previous year and managed to squeeze a little more out of it with a mean of 79 mph and a one-way maximum of 85 mph. They found the other features of the machine much as their opposition and extracts from the test were used in Enfield's advertising.

Motor Cycling had a 250 Clipper complete with Airflow in 1961 and found it good for just over 70 mph with an overall consumption figure of 93 mpg. As the test was carried out early in the year the benefits of the fairing received very favourable note from the rider, while also

Above **1963 version of the 350 cc Bullet which has the Crusader engine, unlike the Clipper which had the Bullet unit. Final confusion**

Left **The 250 cc Super 5 for 1963 with unsprung mudguard mounted to backplate and wheel spindle**

improving the performance a little. It added weight on the front wheel which, if anything, improved braking and it was only at low speeds that it had any effect on handling. In all other ways the machine behaved like the cooking Crusader it was while retaining the traditional Enfield ease of service. It was to be noted that the light switch was not on the left of the speedometer so someone at Enfield read the reports and took notice of comments.

For some time there had been few real changes to the smaller Enfields but, when the range for 1962 was announced, it contained two new versions of the Crusader, one with the exciting features of a five-speed gearbox and leading link front forks. This was the Super 5, quickly dubbed the 'fiver', and it used the basic Crusader parts in the main. The engine was modified to increase its power to 20 bhp at 7500 rpm and this was done by fitting a well domed piston giving a compression ratio of 9·75:1, a bigger Monobloc and a new connecting rod. This was in RR56 alloy and the cap was retained by two cap head screws which were a close fit to hold the alignment and screwed into thread inserts in the rod.

The gearbox was based on the 4 speed design but had three pairs of gears which moved together along the shafts to engage with driving dogs in turn. They were dimensioned to fit into the same space as the 4 speed box and followed the same layout.

The unit was fitted into a standard frame with swinging fork rear suspension but at the front went new leading links based on prototypes tried in 1957 on one of the twins. The forks retained the normal top and bottom crowns supported on head races and these carried two tubes which had rearward extending pressings welded to their lower ends. These carried the fork arms which pivoted on bronze bushes and were controlled by separate Armstrong spring and damper units each of which was concealed in its respective fork tube. The arms extended forward past the unit pickup points and had split ends to

The Crusader Sports for 1963 when it reverted to the normal rear mudguard support in place of the styled one

The same Sports model fitted with the slimmer Sportsflow fairing offered alongside the Airflow. Same year—1963

clamp on to the wheel spindle. They were shrouded externally by plastic covers and the left assembly had provision for anchoring the front brake backplate via a floating link. The brake itself was a 7 in. single sided unit while the rear wheel retained the 6 in. unit with the cush drive and qd facility.

The front forks were fitted with a new pressed steel version of the casquette without the pilot lights, while the front mudguard was fixed to the forks and very deeply valanced so that it carried the number plate on its side. It had a red flash along its centre and was matched at the rear by a styled mudguard that followed the contours of the dualseat as it swept back to the rear number plate. It too had red flashes to emphasise the lines and these were repeated down the side of the fork tubes.

The fuel tank was the usual deep Crusader type and the standard centre section was used and included space for the optional air cleaner. Drop bars supplemented the sporting lines and a rev-counter was available as an extra bolted to the top of the forks. No Airflow fairing was available.

The second new machine was of a totally different type, although very much based on the Crusader. It was the 250 cc Trials model developed from the firm's experience with a works prototype and in stark contrast to the Super 5. It used the Crusader Sports engine with compression ratio of 8·75:1 with a short high level exhaust which ran along the right side of the machine. Ignition was still by a coil which was fitted beneath the seat so the alternator, rectifier and battery were still carried, the last in an abbreviated centre section. No lights were fitted and a bulb horn provided the audible means of approach. A gearbox with a wide set of ratios was used and the unit went into a standard frame.

The front forks were special and contained two-way damping while both wheels were of suitable size and carried the correct type of tyre for the machine's habitat. The sizes were 2·75 × 21 in. front and 4·00 × 18 in. rear. The mudguards were light alloy and the front was unsprung so followed the tyre contour closely.

Fittings included $2\frac{1}{2}$ gallon tank, saddle, air cleaner for the Monobloc carburettor and speedometer mounted in the fork crown. It was a handsome machine but to some eyes lacked the appeal of the Bullet range. It was also rather late in the market place for the trend to the two stroke trials iron was well underway.

Of the existing models both Clippers and the standard Crusader continued without change while the Sports Crusader gained the styled rear mudguard of the Super 5 and a two-tone seat. The earlier chrome plate guard continued to be offered as an option. During the year it gained another option, this being the five-speed gearbox which could be fitted as original equipment for an added £7 10*s* 0*d* on the price.

In July 1962 both magazines road tested the Super 5 and returned very similar top speeds of 83–84 mph. They found that the extra gearbox ratio helped acceleration, while low speed running was of an acceptable nature due in part to the cush drive rear hub and despite the high compression ratio. Fuel consumption around the 70 mpg mark indicated some hard driving and the excellent handling encouraged this, which was not impeded by any lack of cornering clearance.

The low bars restricted the lock a little and, although comfortable, would have been nicer if a little higher. The footrests were in the right place, the seat comfortable, and the front brake very sharp indeed. The gearchange had the usual Albion short travel but comments on the change varied for *Motor Cycle* found it light and positive but did find an occasional false neutral between second and third. *Motor Cycling* had no such problems but felt that the pedal load on the toes taken with the number of gear changes was a little hard and would have benefitted from a lighter action.

Towards the end of the year the single cylinder range was again altered to reduce the number of manufacturing variables, while keeping plenty of options in front of the customer. This was done by eliminating the Crusader which became supplanted by the Clipper, and replacing the 350 Clipper by a New Bullet model based on the Crusader unit.

In addition, the range for 1963 contained an excitingly styled newcomer, the 250 cc Continental, which used the Super 5 engine and gearbox unit in a standard frame fitted with

Right side of the 1963 Crusader Sports

A Clipper from about 1963

telescopic front forks. The appearance was greatly set off by an Italian style 'jelly mould' tank and such touches as an air scoop for the 7 in. front brake, Thruxton drop bars and ball-ended levers. The fuel tank had a snap fastening filler cap and the top fork crown dispensed with the usual casquette and was a light alloy casting that carried the rev-counter, fitted as standard, and speedometer side by side. Just behind the two instruments lay the light switch and they were protected by a perspex flyscreen with the front number plate carried across its base. The headlamp was mounted on brackets from the forks and carried the ammeter in the top of the chrome plated shell. The front end appearance was assisted by a short, chrome plated, front mudguard with a single rear stay and this was matched at the rear by a chrome guard and by the exposed rear unit springs.

The New Bullet looked very much as the Sports Crusader except for an extra couple of fins on the barrel. Internally it replaced the 250's short stroke with the traditional measurements of 70 × 90 mm, but was otherwise strongly based on the Crusader concept. The most important change was to the crankshaft, which was built up with a floating bush big end bearing in the best Bullet fashion and was also given extra support in the form of a plain bronze bush main bearing on the drive side outside of the usual ball race. The gearbox was reinforced to deal with the 22 bhp at 6500 rpm that came from the 7·5:1 compression ratio. For the rest, it looked just like the Crusader Sports in style and finish with the only give-away being the rocker box cover tucked further up under the tank.

Both new models were finished in either a bright polychromatic blaze or gold with extensive chrome plating. This was used on the other singles except for the Trials model and the Clipper. The latter was brightened up with a two-tone finish in flame and cream, while the engine

The 1964 Turbo Twin, essentially a set of Crusader cycle parts propelled by a Villiers 4T engine. Silencers and carburettor from same source

was fitted with the light alloy head and 9·0:1 compression ratio which became common for all four road-going 250s. All models had bonded rubber rear fork pivot bushes which replaced the earlier bronze ones and the sports 250s received new absorption type silencers.

The Crusader Sports and the Super 5 both changed their compression ratios to 9·0:1 and revised their mudguard arrangements. On the first the styled rear guard was replaced by a normal one, and on the Super 5 the front became unsprung and unvalanced. This entailed some special mountings for it, the left side bolting to the brake backplate and the right to a distance piece on the wheel spindle. The Trials model continued unchanged but with the option of lower gearing if required, this being achieved by fitting a smaller engine sprocket.

Also new was a lighter and slimmer fairing, the Sportsflow, of dolphin type with screen, that was available for all singles, while the Airflow continued if wanted on Clipper, Crusader or Bullet.

For the rider who already had a Crusader Sports and craved the looks of the Continental, dealers Deeprose Brothers of Catford, London, produced a kit of parts that transformed the machine. It was not all looks either for inside the engine went a 9·75:1 piston and special cams, while the ports were modified to suit the bigger carburettor that was fitted. The exhaust pipe was swept back and a special end cap with extra baffle was fitted or, for racing, a megaphone could be supplied.

The cycle parts received full attention with clip-ons, rear sets, reversed gear pedal, special rear brake pedal and racing style dualseat. Sports pattern light alloy mudguards were fitted along with fork gaiters, exposed rear fork springs which were chrome plated, and a separate headlamp shell with the same finish. The abridged fork crown as used with the Airflow was fitted but with the speedometer in the mounting and both

681 JNP

three and five-gallon fibreglass fuel tanks were available. Finish was in polychromatic blue and the result a very smart machine.

Early in 1963 *Motor Cycling* tested a Continental with very similar results to that of the Super 5. The substitution of telescopic forks in place of the leading links seemed to make little difference and the cornering limit was when the stand touched down. Even the rather crouched riding position was considered to be well devised and a good example of the type.

1964 brought a contraction to the range in real terms for the number of models was padded out with standard and de-luxe versions of the Crusader Sports and Continental. In both cases this was done by painting mudguards and tank instead of plating them and by a reduction in trim. On the Continental both the rev-counter and the flyscreen were left off. A curious move for the machine sold as much on appearance as performance, and one that Enfield were to reverse inside a year.

Out from the range went the Super 5 and the trials model, although this was still available to special order. With the Super 5 no longer listed the leading link forks went on offer as an option for any of the 250 or 350 cc models.

All models were fitted with a slimmer, longer, fuel tank with slightly reduced capacity and a new silencer designed to keep the noise level within the then impending limits. Otherwise the existing range continued as before with the Clipper becoming available with the five-speed gearbox option.

There was one new model and this broke fresh ground for Enfield as it had a 250 cc twin cylinder two stroke Villiers engine. This was the 4T unit complete with its four-speed gearbox and it slotted into the Crusader frame as if made for it. The rest of the machine was Clipper with Enfield

Geoff Duke stands between a Continental and a Turbo Twin. Geoff was involved with the company's racing plans at that time

telescopics and 6 in. brakes front and rear. The twin silencers were Villiers units, as was the carburettor with its mesh air cleaner. One variation from usual was the fitment of a separate headlamp shell and the whole machine looked very smart in Clipper colours of flame red and cream. It was also very competitively priced at £195 0*s* 0*d*.

A road test followed in *Motor Cycle* which showed it to be a touring machine much as the Clipper with a 70 mph maximum. Consumption was higher than with the four stroke engine and there was far less low down pull, so much more gearchanging was needed. The two were related as effect and cause, and were typical of a two stroke twin needing to be buzzed for performance. In most other respects the machine behaved as the four stroke model with handling being, if anything, better thanks to the lower centre of gravity. All told it was an interesting diversion by the company, which was followed up in March 1964 with a Sports version fitted with dropped bars and chrome plated mudguards and tank panels.

It was apparent that not enough people bought the optional leading link forks or bent those on their Super 5, for the 1965 range included yet another 250 in an attempt to shift them. This was the Olympic which was standard Crusader Sports with four-speed gearbox but fitted with the link forks, the Super 5 pressed steel casquette that went with them, and the styled

rear enclosure. It had dropped bars and a new style to the tank finish common to several models. This was based on a broad stripe which connected into the kneegrip area and was balanced by forward side panels which joined into the top stripe. On the Olympic this was in flamboyant blaze or blue along with the rear mudguard with the stripe and the front forks in silver.

Not much happened to the rest of the range aside from the new tank styling and some colour changes with the Bullet in eggshell blue and white. This, like the Turbo Twin, was unchanged mechanically while the other three four strokes only had a modification to the silencer and the addition of front fork gaiters in the case of the Continental. This was also available in white and chrome as well as the existing colours.

Then, in November 1964, the range was augmented by one further model, the Continental GT or Gran Turismo. It was introduced with a flourish for one machine was flogged from John O'Groats to Lands End in under 24 hours by a team of five riders. Its consumption figure was on the right side of 80 mpg and it fitted in a few quick laps at Silverstone in the hands of short-circuit ace John Cooper.

The GT was the epitome of the mid-1960s café racer and a very real and successful attempt by Enfield to get with the image of the young customer of that time. It was a Continental plus a little more power but a lot more style and line,

Far left **The last word in a café racer from a factory—the 1965 Continental GT**

Left **Engine room of the GT. Note remote gearchange, folding footrest and exposed electrics**

just as demanded by the teenager of the day. It came about because Leo Davenport had taken the trouble to ask his dealers what they wanted to put on their showroom floor and had taken note of their points. His masterstroke was to produce a prototype and then get the enthusiastic apprentices in the works to add their views. The result was a stripped down, race-styled model with all the features the young craved for and would buy. Their fathers would plump for the Crusader Sports as better value but what teenager could resist clip-ons, rear sets, a flyscreen, a swept back and cranked exhaust pipe and much more besides.

The GT had all this centred on the 250 cc Crusader unit with 9·5:1 compression ratio and the five-speed gearbox. As well as the lovely exhaust pipe it had a long bellmouth on the Monobloc without any nonsense with air cleaners or even the option of one. To enhance the racer look there was a massive breather taken from the crankcase just behind the barrel with a large clear plastic tube that ran up under the seat to the rear number plate just as on four stroke racers of the day. The image was continued with the two instruments, as on the standard Continental, and with alloy discs on the front hub. These were commonly fitted to racing singles at the time and, with their row of lightening holes, were much admired and copied. On the racers they had a job but on the Enfield all they did was increase the unsprung weight and look pretty.

The gear lever was connected by a linkage and the well-back footrests and clip-ons augmented by a humped back dualseat. The front forks were gaitered and the rear springs exposed, a small battery fitted behind the engine unit with the ignition coil above it and no toolbox at all. At that time the 'in thing' was to carry the spanners in your jacket pockets—or the lining!

The crowning glory was the fibreglass fuel tank in red to match the mudguards. It was a good size, equipped with a quick action filler cap, had knee recesses and, best of all, 'Royal Enfield' in really large letters on each side. It had no reserve but shouted the name to all around.

In practice it did all the things expected of it with a hard rear end and light front one. You could really feel the road, and over the bumps it would leap about just like a racer. Its saving grace was good handling and stable steering so that it kept its rider on the road even when he overdid matters and dived into turns too quickly. On such occasions the light weight enabled the stock front brake to really calm things down. All this went to mask the rather cooking push rod engine under the tank as the machine was powered out of the corners. In fact, with the wide power band and cush drive, it was very smooth and this was not really in the image.

At the time it was good for 86 mph with the rider prone and returned 76 mpg during a road test. It was also very noisy and this really did

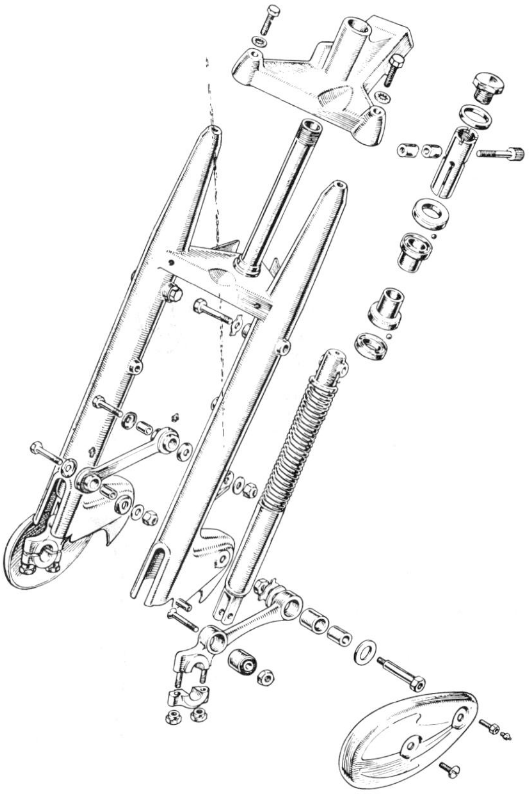

restrict the performance in town. Out of town it was just as bad for it sounded like a racing engine with megaphone so proved to be a magnet to all patrol cars! It was bad for the motorcycling image but enormous fun and just what young riders were looking for.

It might have been what they liked but not enough riders could, or would find the £270 0*s* 0*d* Enfield were asking for it and, while the GT represented the last word in the stripped down café racer style, the Japanese were offering some very fully equipped machines that represented extremely good value for money. With the contracting market of the times sales of the Enfield were insufficient to finance new development and, in August 1965, four models had to be dropped to enable the factory to concentrate on the GT. Out went the standard Continental, the Olympic, the standard Turbo Twin and the 350 cc New Bullet.

The remaining models continued with no change apart from a fibreglass cover over the coil and ignition switch on the GT. At the end of the year the Clipper also went from the range to leave just three 250s, one a two stroke. Rather sadly the company finance problems were to remove these one by one with the Crusader Sports the first to go in June 1966. October saw the end of the Sports Turbo Twin and so, at the end of the year, only the GT remained.

It too was dropped early in 1967 as the company hit the financial rocks and was taken over by Norton Villiers. And so ended the last of the Crusader machines, a far cry from the utilitarian machine of 1956, the most outrageous, noisy and uncomfortable of them all but equally the most enjoyable.

Far left **Exploded drawing of the leading link forks showing the spring units hidden within the legs and the brake torque arm**

Left **A GT fitted with Speedflow fairing with transparent nose. Used to improve penetration but banned from racing**

8 GP5

John Hartle on the GP5 prototype during tests at Oulton Park in 1964

The story of the road racing GP5 begins with the appearance of an Enfield two stroke scrambler late in 1963. This factory prototype used the 250 cc Villiers Starmaker engine in a conventional frame with swinging fork rear suspension. Less usual were leading link front forks and these carried a full width hub. The machine was ridden by Bill Gwynne but lacked the speed of other factory models of that time.

Early in 1964 the road racer itself was first described in the press, although it too was still at the prototype stage. In that form it used the Starmaker engine as prepared by Villiers but fitted with an Enfield expansion chamber. The engine breathed through an Amal GP of $1\frac{1}{2}$ in. bore and the inlet tract was kept short by mounting the GP very close to the cylinder and by leaving off the bellmouth.

The frame was of the usual form with the rear fork controlled by Girling units attached to a rather wide rear sub-frame made that way to accommodate the expansion chamber then in use. The front forks were Reynolds leading link with the spring units fitted into the main tubes to which rearward pressings were welded to carry the link pivots. The handlebars were fitted directly to the fork yoke and located by serrations so they could be adjusted fore and aft but not vertically.

The wheels had 18 in. alloy rims and racing Dunlop tyres, the front built onto an Enfield Super 5 full width hub with single leading shoe brake and small air scoop. At the rear a Motoloy hub was used as this allowed a detachable sprocket to be fitted. The rear brake was cable operated from the left, while a rather uninspired linkage on the right selected the gears. At that time a separate tank and seat were fitted, the first secured at the rear by an aeroelastic and fitted with a quick action filler, the second with a small hump and tail. A fairing was fitted and a rev-counter rather high above the fork crown. The machine weight was given as 210 lb.

This machine was not only a prototype but also an interim model which used the Villiers engine to get the project mobile as quickly as possible. This enabled chassis development to proceed and experience to be gained while the firm got on with work on their own power unit.

Like the Starmaker this was a single cylinder two stroke and used the Alpha bottom half as its starting point. Engine dimensions were 66 × 72 mm and the geometric compression ratio 12·5:1 into a top hat section cylinder head. While the bottom half was very similar to other firm's engines it differed in that the front cylinder studs were more widely spaced and the crank-case halves had transverse cooling fins round the outside. Internally it had the normal Alpha pressed together crankshaft with full circle wheels, oval section connecting rod and caged needle roller big end bearing. The small end had a bronze bush.

The crankshaft was supported by a double

The Enfield scrambler of 1963/64 fitted with the Villiers Starmaker engine. In this case ridden by Bill Gwynne at a TV meeting

row caged roller bearing on the right and a single row plus a ball race for location on the left drive side. The right end carried a cam for the coil ignition points mounted under a cover and drove a rev-counter right angle box from its extremity.

The top half was totally Enfield and air cooled. Both head and barrel were cast in aluminium and sat on four studs with long through nuts which located on a series of diameters as well as clamping down. Production barrels had a cast in iron liner with all ports machined before casting. The prototype used a light alloy sleeve with hard chrome bore which would slip from the finned muff when the engine was cold. This allowed port changes to be easily made and the assembly became gas tight as the engine warmed up. The very precise fit eventually wore and allowed gas to leak but not before the experimental tests were complete.

The barrel porting, like the rest of the engine, was designed by Hermann Meier, an expert in the theory and practice of the two stroke engine. As his brief precluded a rotary inlet valve he used a simple round $1\frac{1}{2}$ in. hole bored into the barrel and inclined at 20 degrees as the inlet port. This carried an Amal GP of the same bore. On the exhaust side the port was bridged and kept very short with a manifold bolted to the barrel to provide a slip joint to the expansion box and anchorage for the two springs which retained it. On the prototype the pipe was offset to the left so much that the left downtube had to be modified to let it past, but production engines had a central port and very little offset.

The transfer ports were in the five port style with the single, large transfer on each side of the barrel splitting into two passages and feeding through two ports. The rear one on each side had a further opening cut through the wall below the port itself and this corresponded to a port window cut in the piston wall above and to the rear of the gudgeon pin. These helped the gas flow from the restricted under piston area and assisted in cooling the small end.

The piston itself carried two Dykes rings, the upper one flush with the piston crown, and was drilled with two small holes in its front on the centre line. These allowed a little oil to reach the very hot port bridge.

Power output was given as 34 bhp at 8000 rpm to 8250 rpm and this was passed via a duplex primary chain to a five-speed Albion gearbox. The box was built as a separate unit but attached at a fixed distance to the crankcase on four studs. Between the two was a cast alloy spacer which was hollow and finned and whose purpose was to allow cooling air to circulate

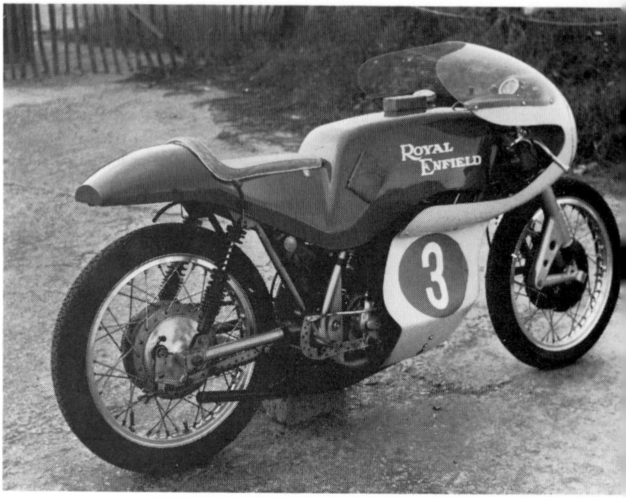

Above **Production model GP5 with combined seat, tank and tail fairing**

Right **Well known picture with Geoff Duke on the early GP5 in the form tried by Hartle**

round the crankcase. The primary transmission was fully enclosed in a typical Enfield chaincase in cast alloy with the other half retained by a single nut and sealing against a round section rubber in a recess. The clutch was multi-plate with six springs, three of which were adjustable for load.

The prototype was first seen at the 1964 Manx where it was ridden by Dennis Craine and managed by Geoff Duke. Normally works machines were barred from this event but, as a prototype, the organising club felt that it could be allowed in. The new engine and gearbox unit was at that time running on a compression ratio of 10:1 and used a smaller carburettor. The cylinder head did have the four horizontal cooling fins merging into five short vertical ones that were to characterise it.

The frame was a full duplex with single top

tube and had the left front down tube cranked to get it round the exhaust pipe. The rear fork had split clamps at the ends for the hub spindle and chain adjustment was by moving the pivot spindle using a range of alloy discs with off-centre spindle holes to locate it. The rear hub was an Oldani with cush drive and the front an Enfield carried in the Reynold leading link forks. Alloy rims were fitted with 2·75 × 18 in. front and 3·25 × 18 in. rear tyres.

Surmounting the machine was a very elegant fuel tank, seat pan and tail fairing all combined into one unit in aluminium alloy. The tank took 7 gallons. Only a top fairing was fitted, this being an Avon unit with transparent nose blister over the front number plate, this being a streamlining tactic tried then to bypass the drag of the flat racing plate. It was soon disallowed.

The machine failed to make any great mark on the race but its performance was good enough to persuade the firm to commence building a small batch for sale. These were to be promoted by a works team managed by Geoff Duke with John Hartle as rider.

Near the end of 1964 a scrambler model joined the Enfield range and was based on the works prototype. Unlike the road racer it retained the Villiers engine and this was equipped with a four-speed gearbox and an Amal Monobloc. Compression ratio was 12:1 and power developed 22 bhp. An air cleaner with two paper filter elements was fitted.

The frame was based on the standard one but had a duplex cradle section bolted to it carrying the engine and gearbox unit. The rear units were laid down a little and the leading link forks retained at the front. Fittings included an alloy fuel tank of $2\frac{1}{2}$ gallon capacity and chrome plated side panels, an overall cream finish, a short competition seat and braced bars.

During this period there was also another offshoot built up for road use to try an Alpha engine. This used the Crusader Sports cycle parts for convenience and in them went the experimental 250 cc single engine with rotary valve and carburettor behind the barrel on a steeply inclined inlet tract. It was a unit that had to be revved to make it perform.

As the start of the 1965 road racing season approached there came news of a stretched works machine near 350 cc which was due to go on test. What was of more interest to customers was the appearance of the first production model GP5 which had received a few alterations over the winter. The engine had the more central exhaust port and the $1\frac{1}{2}$ in. GP carburettor with 'matchbox' float chamber hung on the left. The tank, seat and tail unit had been slightly restyled and were in fibreglass being made by Mitchenall Brothers and finished in red with the name 'Royal Enfield' on the sides in white. And very smart it looked too.

Smart, but never too successful. Certainly there was a potential, for Percy Tait was timed at 132 mph in the 1965 North West 200, but there was nothing else of any moment. Work, however, continued and Neil Kelly came very close to the winner's record lap time in the 1965 Manx before being forced out with clutch trouble. This was to be a frequent story, with finishes far too rare and reliability lacking.

In September 1965 Enfield stopped racing for that year among rumours of quitting altogether, but were just concentrating on the production racers and distilling the lessons learnt from the season's racing. This move did leave Gordon Keith, their front runner, without a machine and Geoff Duke bought one of the bikes for Percy Tait to ride.

Above **At the trials at Oulton Park—some adjustments are made to the early type GP5**

Right **Mick Bowers in the 1964 Experts Grand National on the Enfield scrambler**

The Enfield racing engine became involved with another make in 1965, the DMW and their 500 cc racing twin. At that time Harold Nock of DMW was building such a machine and his first and main prototype used two Starmaker engines joined together. He also built a second machine using two GP5 engines coupled in similar fashion with gear drive to a jackshaft and then to the Albion gearbox. This machine was not taken further and the first twin was overtaken by events and the Yamaha.

Early in 1966 Enfield decided to continue and there were some minor changes incorporated into the engine but it was the middle of the year before the machine appeared with Selwyn Griffiths aboard. The Duke owned model, ridden by Tait, had some outings but success still eluded

ROYAL ENFIELD

them and the company itself was running into trouble.

In the end they were of course taken over by Norton-Villiers and at that time, early in 1967, the whole of the race shop contents were bought by Sunderland dealer Tom Cowie for £4000.

So ended an enterprising attempt to join in the racing world along with a number of other firms

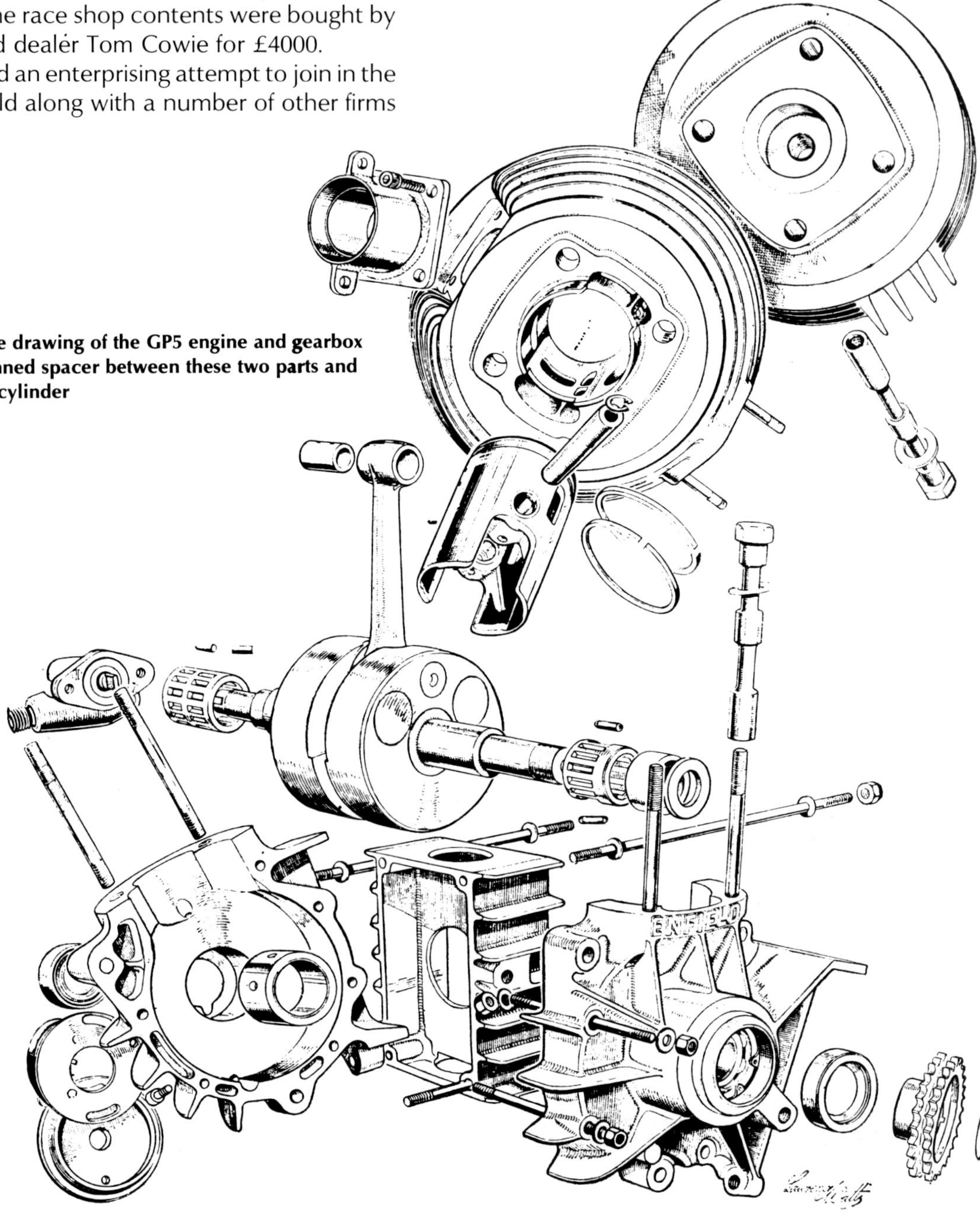

Exploded line drawing of the GP5 engine and gearbox unit. Note finned spacer between these two parts and holes in the cylinder

using the Villiers engine and to progress beyond that with their own power unit. It should have had better luck and more success but this was to go elsewhere before all such attempts were rendered abortive by the TD and finally the TZ.

Figures on the numbers of GP5s built vary from the 100 suggested by the engine and frame numbers in Glass's check book down to about 20. This last is confirmed by the records held by the owners club which indicates frame numbers of 53–59, 61–65, 69–73 and 76–77 to make a total of 19 machines built between April 1965 and March 1966. Some 6 or 7 still exist, one or two still being raced in club events in 1980; which must prove something as the bike was first built as a club racer.

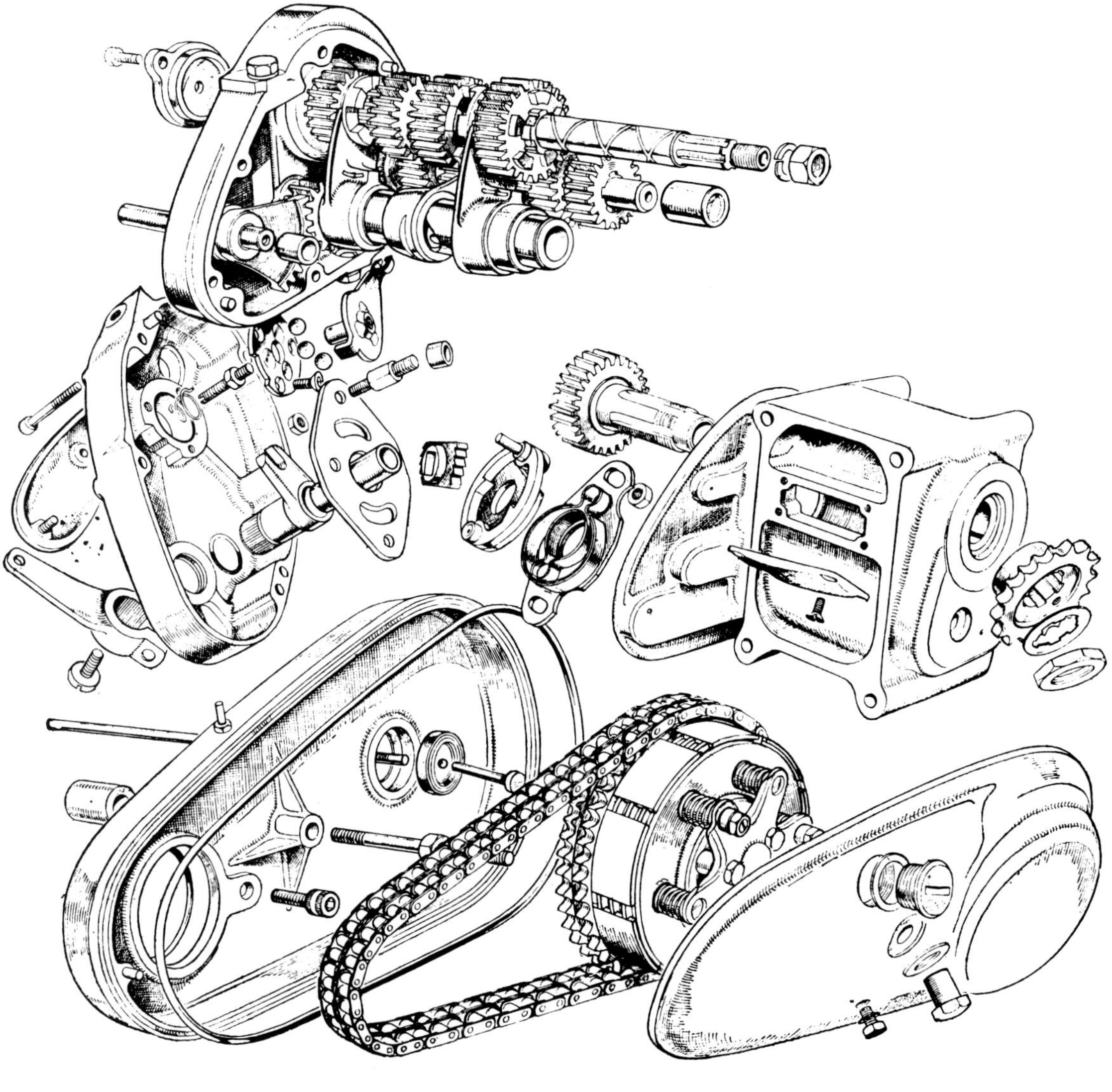

9 Flying Flea

The smallest Royal Enfield had its roots in the pre-war German DKW just as did its Small Heath rival, the Bantam. However, unlike the BSA, the Enfield was developed in 1939 and for very different reasons. The story goes that in the 1930s the 98 cc DKW model RT sold very well in Holland and was imported by a Rotterdam based company. This was Jewish owned and thus in 1938 the importing concession was removed overnight and granted to a firm already dealing with the DKW cars.

The outcome was that the original importers looked to England for a company that would build them a copy of the RT98 and struck a deal with Enfield. The result was an exact copy of the cycle parts but a larger engine of 125 cc in order that it would outpace the German machine. The model was to be called the Royal Baby so that its initials, RB, would be similar to those of its opposition.

In April 1939 two prototypes were on display in Rotterdam and a few machines did reach the public in the autumn of the year around the time of the outbreak of war. This of course left Enfield with a lightweight model fully developed and tooled up for manufacture. During the conflict it was taken up by the army for duties in general and airborne drops in particular. With its special

The 125 as first produced in 1939 with Major F. W. Smith behind it. Both stand outside the company registered offices

CWP 899

Above **The left side of the 1939 model that was soon to join the army for the duration**

Below **The front fork suspension with its rubber bands copied from the DKW RT98 and earlier used for the 1935 works DKW racing 250**

Above **The Flying Flea in its wartime khaki. Note headlight mask, extra silencer and service type kickstart lever**

Right **The Flea in its airborne crate. For this job it has a smaller lamp and no transverse exhaust box**

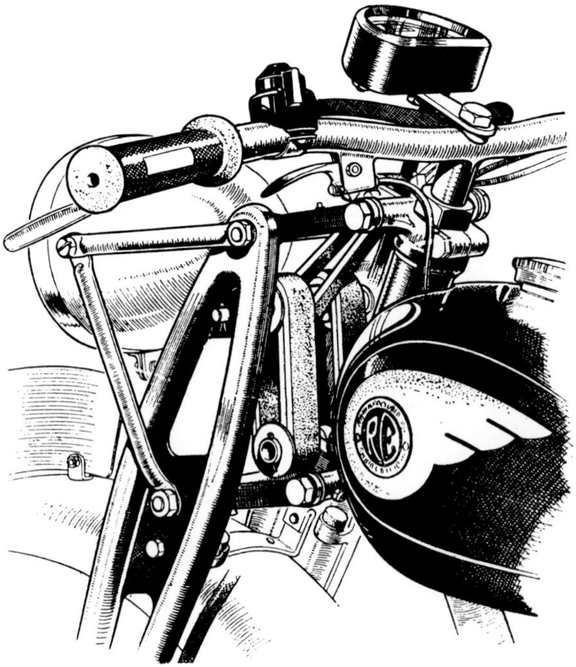

use and light weight it soon gained its nickname the 'Flying Flea'.

Among many other activities Enfield produced over 55,000 motorcycles during the war and a large proportion of them were the model RE, as the Flea was officially known. For the parachute drop it went into a tubular crate, also designed and built by Enfield, that gave it protection and enabled them to be pushed out of aircraft quickly and easily. Smart work on the ground soon had them running and they gave the troops greatly enhanced mobility in comparison with men on foot. Unlike other military machines the Flea was light enough to be manhandled by its rider over obstacles so could be used over most types of terrain.

When the war ended it was an easy decision to continue building Fleas but in civilian guise. It was developed, the tooling existed, and it made excellent utility transport capable of going just about anywhere.

From the start the Flea was of true unit construction with the three speed gearbox in the same vertically split crankcase halves as the crankshaft. The first odd thing about it is that it was, in fact, oversize by 1 cc. The press of the time seemed to go to some lengths to avoid mentioning this but its dimensions were 54 × 55 mm and this gave 126 cc. In some material a bore of 53·79 mm is quoted which works out at just under the 125 cc, but in maintenance features the cylinder bore is given as 2·126 in. and this equals 54 mm. It made no practical difference in the running but could have affected its insurance class.

The crankcase was split on the centre line of the cylinder and cast in light alloy. It carried the crankshaft in two ball races on the left drive side and a ball race and an outrigger bearing on the right, magneto one. Crankshaft seals in the form of reamered bronze bushes outboard of the ball races kept the crankcase pressure up. A bob-weight crankshaft was used, not one with full circle wheels, and it was pressed together with an uncaged roller big end. The rollers used in the postwar engine were longer than the war-time ones and this prevented the crankshafts from being inter-changeable. In either case they ran directly on the crankpin and in the rod eye with thrust washers on each side.

The small end eye of the rod was bushed and the piston attached to it had two pegged rings. The gudgeon pin was retained by circlips. The barrel was cast iron and held to the crankcase by four short studs, a further four in its top face securing the light alloy head. The design followed the pre-war German pattern with a nominally flat top piston and two transfer ports in the opposing side walls of the barrel. Early pistons also had two small deflectors or recesses machined into the crown and top land. The inlet port had a stub to carry the carburettor, while the exhaust was threaded externally for a pipe nut. The geometric

Above **From the 1948/49 period, a model RE fitted with an experimental plunger rear suspension system. In practice only the swinging fork was used**

Top left **The postwar model RE as sold in 1945/47. Hand gearchange, no exhaust pipe bulge**

Left **The 1948 model RE with exhaust bulge and optional legshields**

compression ratio was 5·75:1 and the head carried the plug on the right balanced by a decompressor on the left.

The right side of the crankshaft carried a Miller unit providing ignition and lighting with the rotor inboard of the stator assembly. The outrigger main bearing ball race was outboard of this in the stator which was spigoted to the crankcase and held by three screws in slotted holes which allowed the ignition timing to be varied. Right on the end of the crankshaft went an ignition cam which worked a pair of points mounted on the outside of the stator, with the condenser under a small cover retained by a spring arm. The ignition coil was at the top of the stator with a high tension pickup which connected to a short lead to the plug.

Primary transmission was by a single strand chain to a single plate clutch clamped up by six springs. The clutch inserts were carried in the sprocket which simplified construction but varied the chain line as they wore. The primary chain was enclosed in a case formed by one flat and one dome pressing. The flat sheet went against the left side of the crankcase with the engine sprocket outboard of it and the gearbox one behind it. A suitable hole allowed the mainshaft through to carry the clutch and so the rear wall of the chaincase was formed. The domed outer cover, together with a gasket, completed the job and its fixing screws clamped both halves to the crankcase.

A three-speed gearbox of conventional English layout was used in a separate compartment within the crankcase halves. Only the sleeve gear ran on a ball race, the other shafts relying on bushes and the two middle gears did the selection by moving across the box and engaging with dogs and splines at appropriate moments. A single internal lever acted as a selector by moving both gears and had detent notches cut along one edge. It pivoted in the case and was extended out of it to connect to a rod running up to the hand change lever bolted to the right side

Above **Experimental RE with telescopic front forks and plunger rear suspension. Only the first reached production**

Right **The 1950 RE fitted with telescopic forks mounted on the girder links. Replaced on RE2**

of the petrol tank. This moved in a quadrant which could be swivelled for adjustment so that the marked gear positions and notches corresponded with the gearbox below.

The clutch lever was mounted on the right crankcase side and incorporated a hardened screw adjustor. A small pressing enclosed it together with the kickstarter spring. The kickstarter itself worked within the layshaft first gear and its pedal was splined straight on to the end of the shaft. This differed from the war-time unit which had a folding lever to keep it out of the way on cross-country terrain.

The engine and gearbox unit was mounted in a simple rigid frame. This had a single down tube with two engine plates welded to its lower end to pick up on the front of the unit. A similar arrangement with the saddle tube secured its rear and the two were connected by a top tube and the headstock. Twin seat stays ran back to support the rear wheel on each side giving a very neat structure. Under the gearbox went the footrest and centre stand assembly.

The front forks were blade girders with a rubber band spring medium and layout straight off the 1935 racing DKW models. The sides of the forks were steel pressings and two pairs of rubber bands acted as the forks were compressed, while a single one between them took care of the rebound situation.

Both wheels were spoked, with 2·50 × 19 in. tyres on steel rims. Small drum brakes were fitted, only 4 in. diameter at the front and 5 in. at the rear. They ran on journal bearings which gave an improvement on the cup and cone arrangement used during the war. Simple mudguards covered the wheels and carried the number plates and, at the rear, a carrier. The front wheel was controlled by handlebars secured by a single centre bolt in the style of a bicycle and a war-time requirement to allow the bars to be located fore and aft when in the airborne crate. They carried welded on lever pivots and the fixing bolt was also used for the speedometer bracket when this front wheel driven option was fitted.

The headlamp was hung on stays fixed to the fork pivots and had the light switch mounted in its top and a dry battery for the pilot light in its interior. A single connection at the generator supplied the current to the switch for powering head and tail lights. A dipswitch was fitted to the left bar and a bulb horn added by the dealer to make the machine street legal.

A $1\frac{1}{2}$ gallon fuel tank stored the 24:1 ratio petroil and supplied a small Amal with built-in choke and air cleaner. On the exhaust side the pipe ran stylishly round the outside of the generator and back to a tubular silencer mounted parallel to the lower chainstay on the right. Internally it had three conical baffles with a central straight-through hole to make cleaning easier.

The machine was supplied with a saddle and a tubular toolbox beneath it. It was finished in black with just the tank relieved by silver lining, a silver motif, and a round RE badge screwed to its left side. It weighed 130 lb dry and sold for £48 0*s* 0*d* plus purchase tax. Listed extras were the speedometer and legshields.

They proved to be popular and, with a top

Ten 1950 model RE machines for the Post Office. Fitted with crash bars and legshields but less common than the Bantam

speed of 40–45 mph, were capable of putting 30 miles into the hour. With consumption that never looked like dropping below 100 mpg under the hardest driving they excelled at runabout work and short journeys, especially in town. Many riders took to keeping one as a tender to their larger machine and would use them for such runs as the local post box. With no neec

then for a helmet a round trip such as that in the summer could be done before a four stroke was started and warmed up.

Enfield continued to build them with no changes until 1948 when the exhaust pipe was altered. It gained an expansion chamber very close to the exhaust port which was to characterise the machine for many years. To enable this to be more easily made and fitted, the exhaust port thread was deleted and the pipe clipped into place. The claimed effect was improved mid-range pulling and cleaner two stroking under light loads. Thanks to this the gearing was raised by reducing the rear wheel sprocket teeth by one according to *The Motor Cycle,* and adding one on the gearbox if you read *Motor Cycling*. The arithmetic works best with the first.

Inside the engine the piston deflectors went as they had only been of use in a certain type of barrel. Externally the speedometer bracket was angled so that the instrument could be more easily read, the toolbox was made longer, and the wheel rims, handlebars, saddle springs and contact points cover brightened up with chrome plating.

Enfield built more of them and impressions appeared in the press. Most indicated a marked prejudice against small machines in general and two strokes in particular when the machine was first taken over. As it stood up to the heavy usage and minimal attention without protest opinions swung round sharply. The only real snag was the economy as no-one could remember when it had been filled last and it became easy to run out. The machine seemed especially suited to poor conditions as rain, sleet or snow failed to disturb it and all the testers found it a most stable machine. A formal test gave a maximum of 47 mph and the only carp was the minute front brake, which was rather spongy.

It was not until 1950 that further changes occurred with the introduction of simple telescopic front forks which picked up with the old girder spindle positions. This was done with welded on ears, and transverse tie rods with domed nuts braced the assembly. At the same time the colour was changed to silver grey for all painted items, which included a deeper fuel tank with increased capacity. The tyre section became bigger and the toolbox was fitted with a screw type fastener.

This was only an interim move for the 1951

Top **The model RE2 with redesigned and much smoother engine unit, external ignition coil and new telescopic forks**

Above **Special RE2 with pedals, folding footrests and altered bars for Bermuda. About 1951/52**

Top **The RE2 in 1953 with revised tank finish but in other respects unchanged**

Above **1953 Ensign—150 cc, rear suspension and different toolbox shape. Chainguard and silencer position also differ from 125**

model was a total re-design with new engine unit and frame. Some of the detail items may have been from the earlier unit but for all practical purposes it was a new machine. In time it became known as the RE2, which helped to distinguish it from the earlier version.

The engine was based on the same bore and stroke, had a built up crankshaft with roller bearing big end which ran in four main bearings, and had an iron barrel and alloy head. While this sounded much as before the crankshaft was stiffer and the mains arranged as two pairs on either side of the crank. Thus it was really all new.

The crankshaft did still have bobweights and it turned in mains housed in a vertically split crankcase. Different was the drive side seal which was of the face type with lapped faces held together by a synthetic rubber cushion. On the other side a bronze bush continued to do duty. The crankcase was much as before but its lines were altered to tie in with two outer covers, and all four were pressure die-cast to give thinner sections, less machining and better matching. The barrel was still in cast iron and deeply spigoted into the crankcase but retained in the new engine by four long studs which rose through the head as well. This last item continued to have the sparking plug and decompressor on opposite sides of the combustion chamber. The compression ratio remained at 5·5:1 and the slightly domed piston still carried two rings and was attached to the rod by a gudgeon pin held in place by circlips and running in a small end bush. The cylinder continued with stub exhaust for the clip fitting exhaust pipe with expansion chamber, but the carburettor became flange mounted. It had the float chamber built in and, due to the deep spigoting of the barrel, a hole was machined into the top of the crankcase to accommodate it. A second hole below allowed any petroil overflow out.

The generator was still mounted on the right with the stator outboard and the ignition cam on the end of the crankshaft. The stator coils were mounted in the right outer cover which ran back for the full length of the engine unit and the points were on its outside under a small cover. This last had the letters 'RE' cast into it. The ignition coil itself became an external item clipped to the frame top tube under the rear of the petrol tank.

On the left was another change from before and from convention as the clutch was on the other end of the crankshaft. This came about in part to increase the flywheel effect at the crankshaft, which had been reduced by the lighter weight of the new generator rotor, and to smooth out the power delivery. A further consideration was the small number of teeth desired on what would have been a normal clutch sprocket and the effect this would have had on the clutch design. Clutch inertia problems were dealt with by using a single alloy clutch plate with cork inserts. This was directly attached to the engine sprocket and turned on a bush clamped to the crankshaft by a steel clutch driving plate. This plate was splined to a drum that was free to turn between the engine sprocket and the inserted clutch plate. In the splines also worked the pressure plate which was held in place by a single large circlip in the drum and carried six springs compressed against the drive plate. Thus the springs, by reacting in this way, pushed the pressure plate outwards against the circlip which pulled the drum out on to the clutch plate, and this in turn against the inboard side of the drive plate. By pushing inwards on the pressure plate centre the clutch separated, although once again this did alter the chain line which was dependant on clutch wear. This action to free the clutch was done by a quick thread worm in the outer cover which was turned by a cable running into the upper front of that cover via an adjustor.

The clutch drove a single strand chain which connected to a simple sprocket keyed to the mainshaft of the three speed gearbox. Although much as before with the same internal ratios, the box had been turned round to become a cross-over drive type with the gearbox sprocket on the right. This allowed the primary chaincase to become a simple alloy cover and the rear chain was accommodated by suitable shaping of the rear of the right crankcase half. The mainshaft was supported by a ball race, as was the sleeve gear.

Inside the gearbox the selector design was unchanged, although the actual parts were new. Completely new was a footchange pedal mounted on the right on a shaft that ran across the unit to a positive stop mechanism attached to the outer wall of the left crankcase within the primary drive cover. This connected to the selector via a lever and an adjustable rod, while the gear lever itself was on splines so could be set at the optimum position.

The kickstarter was behind and to the rear of the gear pedal and also completely new in design. The pedal was splined to a shaft that turned in the outer right cover which carried the stator and points at its front end. Its inner end carried a quadrant gear and this turned a gear mounted on a pin outboard on the extended mainshaft end. This gear was pushed inwards by a spring behind it and had a face ratchet which would then match up with a similar ratchet fixed to the end of the main-shaft. Operation of the kickstart thus turned the gearbox shaft and, in turn, the crankshaft. When not in use the ratchets were held apart by a flange on the gear which lay behind the quadrant, and this in turn was positioned by a cam form built into its outer side. Thus when turned it also moved in to allow the gear to move and the ratchets to meet. A small cover enclosed the parts in the outer casting.

The new engine and gearbox unit was fitted into a new frame of the same geometry as the earlier one. It provided a four point mounting for the engine as it was a full loop type with a cradle under the unit to join the saddle and down tubes. It was still rigid and fitted with telescopic front forks but these were revised to do away with the girder type supports and introduce conventional crowns top and bottom. The upper one carried the speedometer and had the handlebars clamped to it, these being very neat with welded on pivot blocks for the light alloy levers, a concealed axial throttle cable, and a similar trigger lever for the decompressor.

The front brake increased in size to 5 in. diameter to match the rear and the tyres remained as before, as did much of the general equipment. Headlamp, mudguards, saddle, tank and toolbox continued, while the silencer was tilted up at the end a little to keep it in line with the altered run of the lower chain stay.

Apart from the external coil the electrical system remained the same but with a very neat connection for the wires from the engine by the points cover. All told the whole machine formed a very neat package with cables and wires run unobtrusively along frame tubes. Finish was silver grey with the name in red on silver panels on the tank sides, and the price was £82 11*s* 0*d*.

Only the nickel shortage affected the RE2 for 1952 and, thanks to this, the rims were painted and lined instead of being plated. *The Motor Cycle* road tested one in March and found it still good for 47 mph as before with a nice gearbox, good silencing, stable handling, and ample braking power. I can confirm all this for I owned an RE2 and it did all this and was an easy starter to boot. The engine speed clutch worked well and the detail design and finish was good.

While the RE2 continued with only a colour change to blue grey and a new tank transfer for 1953, riders seeking a little more power and rear suspension were offered a new model, the 148 cc Ensign. This was based closely on the RE2 but with engine dimensions of 56 × 60 mm, a compression ratio of 6·5:1 and 5·25 bhp at 4500 rpm. A bigger carburettor was fitted but otherwise the engines looked the same. The gearing was raised by fitting a smaller rear sprocket.

The rear suspension was by a simple form of swinging fork that pivoted in a lug brazed to the saddle tube. Its movement was controlled by undamped springs that gave the appearance of plunger suspension. This was in a style used in Europe pre-war and had the springs mounted between frame lugs attached to the chain stays. These springs were screwed into coarse threaded rings formed as part of the rear spindle lugs. With the thread corresponding to the spring pitch they could be wound into place before assembly to the frame and provided both bump and rebound springing. By turning them in the rings the suspension could even be adjusted a little.

The rest of the machine followed the lines of the RE2 but was fitted with valanced mudguards, a new shape of toolbox, deeper chainguard and larger fuel tank. A saddle and rear carrier were still fitted and the finish was in polychromatic copper beech with the new tank transfer. This had the name set in five horizontal lines in similar style to the badge fitted to the larger machines.

This was about the only item on the Ensign to be changed for 1954 when a badge was adopted for the tank, while the 125 was no longer in the

range having been dropped late the previous year after its long service through war and peace.

1955 brought some transmission changes to the Ensign with the internal gearbox ratios being stretched a little and the clutch plate altered to use segment-shaped inserts. On the outside, covers were added to shroud the rear suspension springs. The finish became maroon with the options of olive green or polychromatic silver grey also available.

In the new model changes announced in September the Ensign was described as having much wider brakes, larger fork sliders and cover tubes, and some more clutch modifications. In fact, it was superseded in November at the show by the Ensign II which finally broke away from the ultra-simple appearance the model had exuded since its Flying Flea days. The changes to the machine were small but had a very real effect on the looks with only the bulb horn on the bars a relic from earlier days. For the engine there was a new head and barrel still in the same materials but with much more substantial cooling fins. The holes in the head were moved round to lie on the fore and aft axis with the plug laid back to the

Above **The Ensign rear suspension with plunger type springs controlling swinging fork. Limited load adjustment but no damping**

Right **The suspension springs after they had gained covers, painted first and later chrome plated**

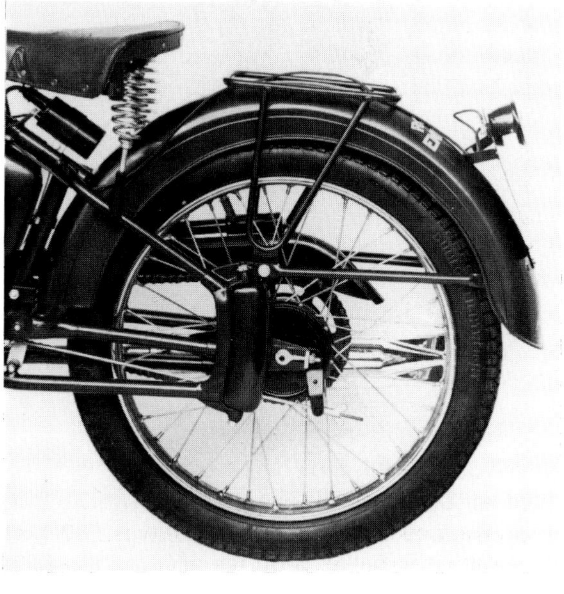

The Ensign II in 1958. The dualseat has a marked effect on the appearance of the machine. Revised plug position in cylinder head

rear and the decompressor forward and hidden among the fins. Inside, the porting and the combustion chamber were modified to improve the performance.

The cycle parts were mainly as before but the lines were completely changed by the adoption of a dualseat. The coil was re-positioned under it and these changes gave a much more solid look to the whole machine, which was finished in surf green with the new style tank badge. A detail change was the use of chrome plating on the rear spring covers.

The Motor Cycle tested one in March, 1956, and found it good for 51 mph while retaining the attributes of its predecessors. Consumption was still close to the 100 mpg mark, the wider brakes worked well, and the machine continued to fulfil its duties just as its forbears had done.

For 1957 the Ensign II was offered in an alternative build with a battery and rectifier for an extra £6 0*s* 0*d* and with this came an electric horn, stop light and ammeter in the headlamp shell. The battery was mounted on the right of the toolbox with the horn to the rear of it. The finish was unchanged and *Motor Cycling* tested one to 52 mph in top at 4800 rpm but in second ran the engine up to 6500 rpm to achieve 41 mph. This did seem to confirm that top gear was too high and that a four-speed box would have been an advantage. Over the test 125 mpg was obtained in mid-winter conditions and they confirmed the earlier findings.

In 1958 the model with rectifier and battery became the Ensign III although the II continued with it. The newer machine had a cast alloy fork top crown which had ears to support a new and longer headlamp in which was fitted the speedometer in addition to the lighting switch and ammeter. Both machines were available finished in the choice of surf green, as before, or black.

The Ensign II was dropped during the year but the 1959 range still included two 150s as the Ensign III was accompanied by a more modern version called the Prince. The engine unit of this used much from the Ensign but changes included full circle flywheels for the first time and no decompressor. The power output was up and in the gearbox the tooth form was improved and the shafts made sturdier. A rubber cush drive was incorporated in the mainshaft sprocket but the gear ratios were unaltered.

This unit went into a new cradle frame with swinging fork rear suspension controlled by conventional Armstrong units. The front forks came from the Ensign, as did the headlight, but both mudguards were new and more deeply valanced than before. A very extensive rear chaincase was fitted that nearly fully enclosed the chain, and side covers closed the compartments below the dualseat. Both wheels had full width hubs and a larger 3 gallon fuel tank was fitted. Finish on the Prince was in cherry red with black frame and forks, while the Ensign continued in surf green. Both became available with the Airflow fairing but the Ensign was dropped from the range during the year.

The smallest Enfield was nearing the end of its days as the accent swung to bigger and more sporting machines. Most first-time buyers went for the 250 cc models or a secondhand machine as the price of the Prince would buy a nice two year old 250. Enfield continued to list it into the 1960s without change and for 1962 it was presented in a two-tone burgundy and cream finish.

This could not save it and in August 1962 the model left the range and so brought to an end a machine run of over 20 years that had begun with a German foible and Dutch determination.

Last of the line, the Prince for 1962 (its final year) fitted with its own version of the Airflow fairing

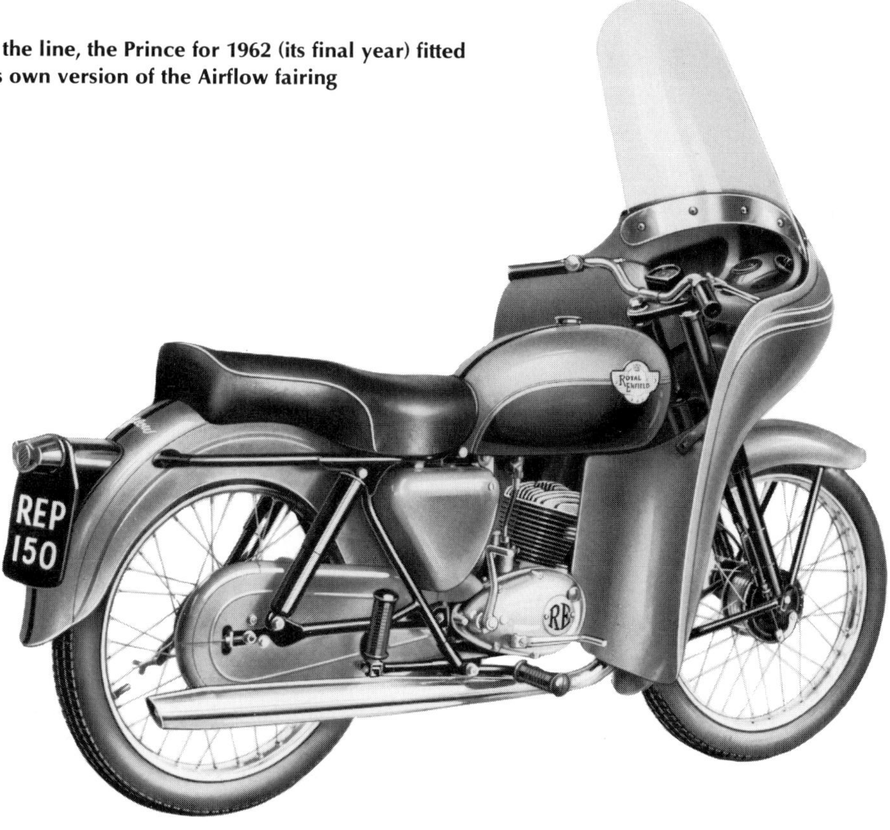

10 Indian Enfield

Early in 1977 the 350 cc Bullet reappeared on the English market in a strange 'coals-to-Newcastle' manner brought about by a move made by the company two decades earlier. It was a step back in time for the new models were to all intents and purposes replicas of the mid-50s Bullet in looks and performance.

What had occurred in the past was that in 1956 Enfield had set up a subsidiary in Madras in India to make the Bullet under licence in its 350 cc form. The factory out in Asia did no more than to reproduce the machine in every detail in accordance with the works drawings and no doubt under guidance from the production men at Redditch. Once underway they were left much to their own devices, due to the distance involved, and while the home machine received its annual crop of changes, out in Madras they just kept to the 1955 specification.

In the end the English Bullet went the way of all the traditional British singles and eventually the firm, the way of the industry. Out in India they just continued to build the same machine and only changed it when forced to. Thus in time the magneto was replaced by a distributor and the dynamo by an alternator, but the electrical system stayed on 6 volts and there was no move to 12 or such things as zener diode control.

That apart, the Indian machine remained unaltered as if frozen in a time warp for all the British riders who sighed for the return of the traditional single. Then in 1977 their bluff was called when the Slater brothers, who already imported the Italian Laverda machines, added the Enfield India to their lists.

When they arrived they proved to be just what had been promised, 1955 Bullets with AC electrics. The engine was just as before with its alloy head and iron barrel. The compression ratio was 6·5:1 and the crankcase carried the oil for the dry sump system in a separate compartment behind the flywheels. The flywheels were of a size not seen for a generation, massive and built up with the expected plain bearing big end bush. They were the centre point on which the machine turned for it was they which gave it its style of leisure and easy rolling uphill and down dale, always in top gear.

The Crusader assembly line of Enfield India in Madras. The machines are powered by locally-made 175 cc Villiers engines

The timing side with its highly distinctive shape embracing oil pump, filter and train of gears was just as before. Only at the end of the train was there a change with an alloy casting housing the drive shaft from the gear to the points cam and its auto-advance all under a small cover.

The top half of the engine was its ageless self with the oil feed to the rockers from the crankcase just as always. Upstairs went the two separate covers for the rockers and lower down the inspection plate for the tappets. The exhaust pipe was low level and the silencer parallel to the ground without the uptilted end of the earlier models. On the intake side went a carburettor that was clip fitted to an adaptor bolted to the cylinder head flange. The instrument was of similar appearance to the Amal Concentric with central float bowl with tickler on right. It was fed from the traditional form of air cleaner box mounted on the right and carrying an instruction to clean the filter every 3000 kilometres. As in the past, the carburettor was fitted with an air slide and this was worked by a lever on the right bars. Equally common in the past was the lever on the left which lifted the decompressor valve in the head on the right opposite the plug. In practice the low compression ratio made it rather superfluous.

On the left of the engine the primary transmission was by duplex chain in an alloy chaincase whose outer half was secured by a

Left **The Enfield Bullet as first imported into the UK by the Slater brothers in 1977. Carburettor is common with 175 so too small here**

Below **Neale Shilton and Edward Turner discuss the prototype police model shown in 1972. Based on the Indian Crusader it was too expensive**

Bottom **Indian Enfield Bullets for the army in 1980. Strange seat bracket to retain passenger**

single nut. The alternator was housed in its front end and the clutch at the rear. The four-speed gearbox was bolted to the back of the crankcase and its right cover carried the neutral finder and concentric gear change and kickstart pedals. All just as before down to the bend in the gear pedal to enable it to miss the footrest, and the oil filler cap sandwiched in between engine and gearbox.

The cycle parts continued this theme with single top and down tubes for the open diamond frame held together by the engine unit. A simple rear subframe supported the rear fork pivot and the top ends of the Indian made, non-adjustable Armstrong units which controlled it. At the front went Enfield-type telescopic forks surmounted by the casquette complete with its two miniature pilot lamps.

Both hubs housed the 6 in. diameter single sided Enfield brake and were spoked into steel rims carrying 3·25 × 19 in. tyres. Tyres which contained a fair, if not generous, proportion of natural rubber which was good for a long life but less so for grip both wet or dry. The mudguards were both deeply valanced and chrome plated, the front one with black registration number patches for use in India.

The fuel tank had a central filler and chrome plated side panels to embellish it and carry the Enfield India badges. Behind it sat a very substantial dualseat with rear handrail attached to it. Beneath the seat, on the left, hung the black old fashioned battery, and on each side in the subframe angle was tucked a good sized toolbox, that on the right housing the ignition key switch and on the left the rear brake-operated stop lamp switch.

The lights were controlled by an old type black switch on the left of the casquette top and this was matched by an ammeter on the right, both items sitting behind a 160 km/h speedometer, a rather optimistic fitting. Turn indicators were fitted as a concession to modern times but nothing else, no warning lights for neutral, oil, or main beam as had become normal by 1977.

In June 1977 *Motorcycle Sport* published a road test on the Bullet which put over the feel of the machine, its low engine speed, and limited power output well. It cruised along at an indicated 100 (km/h) or so, with just a flatter tone creeping into the exhaust note as the engine dealt with main road gradients. In villages, the tester thrilled to the noise of the exhaust rebounding off the walls, which was a good way of summing up Indian noise legislation.

The engine started easily and ticked over with a slow even beat hot or cold, acceleration was sedate, but the brakes were quite inadequate. The electrics took the tester back two decades to the days when Joe Lucas was known as the Prince of Darkness, and not without reason. The machine would run up to 70mph and returned 80mpg which, with the fair sized tank, gave it a good range.

The finish varied with the paint very good on the black model ridden but less so on red and blue ones, some machines having to be refinished by the importers. The chrome was to a much lower standard and on many items the nickel base could be seen. However, the exhaust pipe remained pristine and did not blue near the port.

A little later, in January 1978, *Motor Cycle Mechanics* also tested the Bullet and were less impressed by it, concluding that it was a nice machine but really ought to be in a museum. They had some troubles, but thought it a solid piece of workmanship despite the snags they struck. They did question who would buy such a machine as a sophisticated Japanese 400 was little extra money, while a rider wanting an old British machine would buy just that and not a new old British machine. The Enfield's speed and fuel economy could be had from a small Honda at half the price.

In the end it worked out well enough for the importers as enough people bought the machine to encourage them to continue bringing it into the country. That is until things began to go wrong at the Asian end, so that the supply of bikes dried up late in 1977.

The trouble started with a four month factory closure due to a productivity dispute. Production was resumed at Easter 1978 and output then soon climbed from 1200 to 1500 units per month. This, however, was only enough to fulfil the new orders coming into the factory and to meet these all back orders and export were held in abeyance indefinitely.

Then nature took a hand with a tidal wave and floods which caused devastation, after which there were dock strikes in India. Finally all these problems were overcome and production was increased to cope with all the new and outstanding orders but this had taken most of the year. It was mid-December when the first of what were to be regular monthly shipments left Madras to bring the machines to England to reach the showrooms by mid-January 1979.

From then on matters ran much more smoothly and sales gradually built up. As might be expected there were no mechanical changes but 1981 did bring a revised finish in black with gold lining to the tank. The chrome mudguards and tank panels went, along with the tank badge which was replaced by a simple one word 'Enfield' transfer. The only noticeable variation in specification came late in 1981 when the indicators were deleted.

For the early part of the 1980s, the Enfield simply continued to be available, a curiosity in the lists of new machines. Importation into Britain stopped around 1985, but the factory in Madras continued and made improvements to secure Indian police and army contracts. In 1987 the model came back to Britain, imported by Bavanar Products, with better paint and 12-volt electrics, but otherwise much as in the past.

The range was expanded for 1988, the Bullet becoming available in three forms. These were Standard, De luxe with more chrome-plating, and the Superstar with the tank panels and toolbox in red. All versions had a 7 in. front brake, folding kickstart pedal, turn indicators, better finish and a whole host of detail improvements to overcome legislative needs and to deal with owners' complaints.

The Enfield India engine in 1981 showing the points housing and general mid-1950s Bullet design

Proposed for 1988, but not seen until late-1989, was a 499 cc version of the Bullet, this following the same lines as the smaller model, but being fitted with a single disc front brake. Another year passed before the larger Bullet reached Europe, in part due to delays caused by incorporating changes included to please British customers. One alteration from the machine seen at the launch was a new twin-leading-shoe front brake, the disc being an option to follow later. Also to come was electric starting, mentioned before, but still under development at the factory.

Two mopeds, the Explorer and the Silver Plus, were listed by the firm, these being conventional models with 50 cc two-stroke engines, but with only three speeds in the gearbox. Their general construction and small drum brakes mirrored Japanese models of a decade or more earlier but the prices were low.

By 1991, the larger Bullet was available as a Superstar model, as well as the Standard, while the three smaller versions and two mopeds continued. Little was altered, but quality control in India was tightened up for 1992 as the Enfield India rolled on, remaining much as the 1955 machine it was.

11 Competition

The Enfield competition story is a composite one and the parts are not much related to one another. The section that may symbolise the make and its successes for any one person vary far more than for other machines. After all, fan or not of Nortons, few could associate them other than with road racing and the TT. Not so with Enfield for the name can mean racing, scrambles, trials or ISDT, and the racing can separate into sections some two decades apart.

No attempt has been made to list all the make's successes, but instead some of the highlights are recorded from each area of activity. For a firm with a solid, heavy reputation they can be seen to have had their moments and successes in a broad field of events.

International Six Days Trials

While it may not be easy for an outsider to pick out a competition area that symbolises Royal Enfield, the factory itself had no doubts at all in the post-war years—it was the prestigious ISDT that was the pinnacle that had to be climbed and any other activities were pushed aside for it. From early days they had limited interest in road racing and had concentrated on trials events in the 1930s with considerable success. This effort was continued after the war and to Enfield's way of thinking success in trials could only be bettered by success in six days events. For an exporting company wishing to advertise its products round the world this led to just one

Preparing one of the team 500 Twins for the 1951 ISDT. Lots of detail work was done to speed up repairs

event—the International Six Days Trial in which a team of five riders from a country would compete for the Trophy, and a smaller team of three aim for the Vase. Trophy men had to be on machines made in their own country and competition for a team place was by selection and very stiff indeed. In addition to the main team awards there were others given to successful manufacturers' teams, and in all these categories the riders had to maintain a faster speed schedule than the bulk of the entry. Any rider who completed the trial without loss of marks gained a gold medal which could be lost by failing to clock in on time at any check, or just by not starting up within a time limit each morning.

The Enfield riders first rode for their country in the 1948 event held in Italy with San Remo the centre. This was the second postwar trial, the first having been run with a limited entry in Czechoslovakia the year before. 1948 was very different with seven Trophy teams and 15 Vase ones, and was to be the start of a period of English dominance, for both major trophies were won by their teams. In the Trophy team there were two 350 cc Bullets ridden by Charlie Rogers and Vic Brittain, while Jack Stocker rode a 500 in the Vase. This was a machine he had developed from a model J as he preferred a larger capacity and had earlier declined to ride on the 350. It was a quick machine and Stocker kept it on time despite a fracture in the front forks which he lashed together. Out of line they may have been but they held through the last two days and the speed test in which Jack had to fight the model round every bend. He got a great ovation at the prize giving when the team collected the Silver Vase.

Such tenacity was the very essence of six days riding so it was not surprising that Jolly Jack was retained in the Vase team for 1949, when the event was held in Wales. Charlie Rogers was in the Trophy team which again won, while the

Vase team lost no marks but had to give best to a Czech team of 125s in the speed test which favoured small engines. Enfield returned home with a manufacturers team prize and a total of 5 golds had been won on their models.

1950 was mainly a repeat, again in Wales, again with Britain winning the Trophy and this time the Vase as well. Rogers had followed Vic Brittain's example both having been pre-war experts winners, and retired, to the preparation and testing of Enfield's competition machines. This let Jack Stocker into the Trophy team on a Bullet and Tom Ellis into the second Vase team, but he was unlucky enough to retire with a holed piston in his 500.

Britain won the trophy for the fourth year in succession in 1951, their 15th success since the event began in 1913, with Jack Stocker on a 500 twin. There were no Enfields in the British Vase teams but all three members of the Irish squad and two of the Swedish one used the make which once more won one of the manufacturers team awards. A total of seven Golds were won, four on 500 twins, and one of the successful riders was Johnny Brittain, son of Vic, and destined to be the outstanding Enfield trials rider in the postwar period.

For 1952 Jack Stocker went further into big machines by riding the 700 cc Meteor in the Trophy team. It was a disastrous year for the team but Jack once again brought back a gold. His factory colleagues riding in the Vase teams were less fortunate as both lost marks one morning when their twins refused to fire. Then Don Evans was forced out when his machine caught fire and was burnt out due to a split tank after earlier leaking problems. Johnny Brittain was also forced out when his engine clanked to a halt.

The next year, 1953, was better and Britain won the Trophy once again with the help of both Stocker and Brittain on 500 twins. The Vase team was again beaten by the speed test and Don Evans was one of its members to win a gold.

Above **A very young Johnny Brittain dealing with Stoney Brae in the Scottish Six Days**

Left **Younger brother Pat Brittain on Perwydd in the Welsh Trophy Trial**

Another makers team prize was taken back to Redditch, making three in five years.

That was to be the end of the winning for Britain, although they came close the next year when Stocker and Brittain were again in the Trophy team but it lost out in the speed test. On the last day Stocker went through a hedge and seven feet down into a field. He was fortunate to escape with cuts, but poor Don Evans, riding in the Vase team, made the same mistake and bent his bike too much to continue.

Only Brittain was in the Trophy team in 1955 and, while the team was out of contention, he won his gold. The great days for the English team were past and it was the Czech and East German men who were to dominate the event. Johnny Brittain was still there in 1958, by which time teams were six for the Trophy and four for the Vase, and again won a gold, as did Peter Stirland in one of the Vase teams. Enfield once again won a makers team award.

In 1959 Terry Cheshire was the Enfield Trophy man, and in 1960 it was Peter Fletcher who won a gold that year. For 1961 Johnny Brittain was back in the Trophy team with Fletcher in the Vase and both won golds, although the teams were less successful. 1962 saw both of them in the Vase team winning golds with their 250s, and Johnny went on to another gold in 1963 and yet one more in 1964.

That was the end of the Enfield ISDT effort for the factory was in trouble and the days of the big single were long gone. They could look back with pride, however, to the days when their country had monopolised the awards and their machines had always been part of the team.

Above **Peter Stirland in 1961; note the registration number, Enfield's reply to machine '350 BSA' ridden by Brian Martin**

Right **G. F. Broadbent with a touch of wheelspin in the 1956 Ilkley Grand National**

Trials

As well as the ISDT there were lots of other trials for Enfield to aim their men at, with highlights in the Scottish Six Days, the British Experts, and some other 2 or 3 day events such as the Welsh. They had had some very good years just before the war with one of the best being 1937 when Charlie Rogers won the British Experts and post-war he was soon back in the groove. Teamed with him were J. J. Booker on a 250 and L. G. Holdsworth on a 500. Charlie's first post-war event was the 1946 Exmoor which he won without losing a mark with best time in the tie-breaker. He also won the John Douglas with his team-mates taking the class awards and so established Enfield in the 1940s trials world. In due course he retired to prepare the competition models and his place was taken by J. V. Brittain, Johnny son of Vic the 1936 British Expert.

Johnny Brittain was the mainstay of the Enfield trials team through the 1950s and beyond, as well as performing to gold standard in the ISDT with monotonous regularity. His abilities in one day trials were very quickly highlighted and as early as 1950 he rode in the British Experts. Two years later he won it a month before his 21st birthday. This made him the youngest ever winner and the first to follow in his father's footsteps. That year he also won the Scottish Six Days and was third in the Trials Star for the year.

In 1953 he repeated his success in the Experts and his third place in the Trials Star but dropped to third in the Scottish by a matter of five marks. He continued this level of performance through the 1950s with innumerable trials wins, good placings in the Star and the Experts, and wins in the Scott in 1955 and 1956. By then he had been joined by his brother, Pat, who was third in the Experts that year while Johnny took the Trials Star.

He won the Scottish again in 1957 and that year was part of a British team which cleaned up in the Polish Tatra event. Peter Stirland joined the factory team and they tied for fifth spot in the Experts with Johnny then taking third in the Northern Experts, while Peter won the Southern. Most times the faithful 350 cc Bullet was used, extensively lightened over the years, but on occasion a 500 cc model would be wheeled out, and in the late 1950s all the riders began to use a 250 based on the Crusader model. This was never as successful and Peter Fletcher kept to a 500 as far as he could, still riding the big single in the mid-1960s and usually winning the over 350 cup for his pains.

Despite the presence of some very good riders on other makes, such as Viney, Smith and Miller, Enfield took a good proportion of the trials honours in the postwar era and the main credit for this has to go to Johnny Brittain.

Scrambles

Despite their successes in the ISDT and the Scott, Enfield had little to do with post-war scrambles as a factory. There were Enfields entered in races

Right **Peter Fletcher on his Enfield single in the 1965 Scottish Six Days**

Below **A trials model with the Crusader power unit ridden by Peter Gaunt. The wheelspin indicates a problem**

Bottom **The scrambles Crusader fitted with the leading link forks used for other competition models. Traditional, and clean**

98 DN

Left **An Enfield in the Island approaching Ballacraine. Marked as Bill Lomas in the Manx. He was 7th in the 1948 250**

Right **Bill Lomas at Cadwell Park on his very special 250 cc Royal Enfield, home-converted to twin overhead camshafts**

and one or two did well in some of the better events but the factory kept to its trials. Until, that is, 1963 when a works machine began to appear in TV scrambles ridden by Peter Fletcher.

The bike used was based on the Crusader engine and gearbox unit with a 10:1 compression ratio, energy transfer ignition and a straight-through exhaust pipe. The frame was a standard one strengthened at the rear fork pivot and fitted with leading link front forks controlled by Armstrong spring and damper units. The cycle parts were amended to suit the machine's use and it began to run in events.

Not for long, however, for soon the engine unit was replaced by a Villiers Starmaker and the model described in an earlier chapter came into being. This became part of the GP5 story and, like the road racer, went when the firm was taken over.

Road Racing

Even less is Enfield associated with road racing but they did dabble in it over the years and on occasion their machines were used by private owners to good effect. They ran in the TT as early as 1911, when they were 5th in the Junior, and in 1914 F. J. Walker led at the end of the first lap on his 350 cc vee twin. He had many falls in the race but in the end finished 3rd only to crash after the finish and to succumb to his injuries.

A few Enfields finished in some good places in the TT through the 1920s and in 1928 Cecil Barrow gained a second in the Lightweight, albeit 17 minutes behind the winner and a mere 25 seconds in front of the next man. In 1935 he had his last ride in the TT and returned to the Enfield marque he had used in the 1920s. He rode a 500 cc four valve Bullet in the Senior and brought it home in 8th place.

After the war the Bullet model appeared in the Clubmans TT but was never up with the leaders. Far more successful was a 250 cc special built up by Bill Lomas, destined to become a works rider for several leading makes and a world champion. Lomas was an Enfield apprentice and was racing JAP engined specials with considerable success in the 1940s. His 250 special was based on a 1938 Enfield to which he fitted twin overhead camshafts driven by a single chain that ran in a T-shaped timing chest. Lubrication was extensively modified to look after the new top end and the machine won on its first outing. After Lomas moved on to bigger things the 250 continued in use with other riders until 1955 when the piston melted. It was put to one side for over a quarter of a century but was then brought out for a possible rebuild, but only if the old castor oil stains could be removed from the crankcases—no easy task.

The next Enfield involvement in racing came with the introduction of the Constellation, five of which were entered in the 1958 Thruxton 500 mile event. At first things looked well with Bob McIntyre going straight into the lead, but it was not to be and a tank change and a misfire put them back a place. Poor Bob had some long pushes before he could persuade the reluctant engine to fire after stops but still lapped much faster than the field. In the end Enfields finished second and third.

For the next three years McIntyre tried to get the Enfield home at Thruxton but was out of luck each time despite leading from the flag. Encouraged by this one or two people campaigned them in production events but with limited success. Late in the 1960s the Series 2 was also to be seen in such events, but again was rather overwhelmed by sheer force of numbers. Two Rickman Enfields ran in the 1970 Bold-Or but went out with gearbox trouble.

And so to 1977 when number 105 began to appear at the front of the 350 cc class at vintage race meetings. The machine was a 1950 Bullet and the rider Steve Linsdell who set about the exotic Manx, KTT and 7R camshaft machines along with the occasional Gold Star to such

Left **J. Goldsmith gets the front end well up in a meeting in California. Machine is the 700 cc twin and he seems worried**

Left Below **The camshaft drive of the Lomas 250 cc single. The machine's success reflects Bill's riding and engineering ability**

effect that he quickly became the man to beat. Much to the chagrin of the camshaft brigade Steve's rewards came from little money and a considerable amount of hard work. Internally the engine was tuned up and the lubrication system well modified to ensure reliability. The complete machine was light and Steve rode it hard with the inevitable results, and these continued in 1978, 1979 and 1980. In 1979 he produced a second machine which was a 1953 Meteor prepared to the same standard. As the big Enfield twin was always quick Steve concentrated on keeping the machine in one piece and it proved to be just as successful as the 350.

Not content with the effect his two Enfields had in vintage racing he went on to build a third machine of 500 cc. This was rather more modified than the others for apart from the work on the engine, this item was housed in a Seeley frame and drove a Norton gearbox fitted with five speeds. The machine weighed 290 lb and was entered for the 1981 Manx Newcomers' race. In this it was geared for 130 mph at 7000 rpm and proved completely reliable, requiring only a check over and a change of oil in some 16 laps of practice and race. Steve hoped to lap at 88 mph, the maximum speed of the stock 500, but in fact managed 95·64 and averaged 94·67 to finish second to a Yamaha 4 and in front of a field of modern machines.

Quite a feat for a marque not associated with racing, the TT or even very often with competition success.

Left **Steve Linsdell on his larger Enfield, the 700 cc twin. A very successful rider in vintage racing with this and a 350 cc Bullet**

Below **The engine and gearbox of the 350 cc Bullet raced by Steve Linsdell. The important parts are inside**

Appendix

Specifications

Model	**C**	**C0**	**G**	**J & J2**
Year from	**1940**	**1940**	**1945**	**1945**
Year to	**1945**	**1945**	**1954**	**1955**
Bore (mm)	70	70	70	84
Stroke (mm)	90	90	90	90
Capacity (cc)	346	346	346	499
Compression ratio (to 1)	5·0	5·75	5·75 **1**	5·5
Valve position	side	ohv	ohv	ohv
inlet opens BTDC	30	30	30	30
inlet closes ABDC	60	60	60	60
exhaust opens BBDC	75	75	75	75
exhaust closes ATDC	35	35	35	35
Valve clearance (cold) inlet (in.)	0·004	0·002	0·002	0·002
Valve clearance (cold) exhaust (in.)	0·008	0·004	0·004	0·004
Ignition timing (in.)	0·375	0·375	0·375	0·312
Points gap (in.)	0·012	0·012	0·012	0·012
Primary drive chain	$\frac{1}{2} \times \frac{5}{16}$	$\frac{1}{2} \times \frac{5}{16}$	$\frac{1}{2} \times \frac{5}{16}$	$\frac{1}{2} \times \frac{5}{16}$
Rear chain	$\frac{5}{8} \times \frac{3}{8}$	$\frac{5}{8} \times \frac{3}{8}$	$\frac{5}{8} \times \frac{3}{8}$	$\frac{5}{8} \times \frac{3}{8}$
O/A ratio: top	5·95	5·65	5·5 **2**	5·0 **3**
O/A ratio: 3rd	8·05	7·90 **4**	7·2 **2**	6·5 **3**
O/A ratio: 2nd	10·70	11·2 **4**	9·9 **2**	9·0 **3**
O/A ratio: 1st	15·81	18·6 **4**	15·3 **2**	13·8 **3**
Front tyre (in.)			3·25 × 19	3·25 × 19
Rear tyre (in.)			3·25 × 19	3·50 × 19
Brake front dia. (in.)			6	6
Brake front width (in.)			1	1
Brake rear dia. (in.)			6	6
Brake rear width (in.)			1	1
Front suspension	girders	girders	teles	teles
Front movement (in.)			6	6
Rear type	rigid	rigid	rigid	rigid
Petrol Tank (Imp. gal.)			2·75 **5**	2·75 **5**
Oil tank (Imp. pint)			4·0	4·0
Box capacity (Imp. pint)	oil/grease	oil/grease		
Ignition system	magneto	magneto	magneto	magneto

Model	C	C0	G	J & J2
Year from	**1940**	**1940**	**1945**	**1945**
Year to	**1945**	**1945**	**1954**	**1955**
Generator type	dynamo	dynamo	dynamo	dynamo
Output (watts)			35 **6**	35 **6**
Battery (volt)	6	6	6	6
Wheelbase (in.)			53·5	53·5
Ground clearance (in.)			5·0	5·0
Seat height (in.)			29	29
Dry weight (lb)			358	
Wet weight (lb)			381	
Power: bhp			15	21
@ rpm			5400	5000

1 1951–6·5 **2** 1948–5·6, 7·3, 10·1, 15·6 **3** 1948–5·1, 6·6, 9·2, 14·2 **4** Burman ratios–7·23, 11·75, 17·85
5 1947–3·75 option **6** 1950–60

Model: **Bullet**	**G2**	**500**	**India**	**500 India**
Year from	**1948**	**1952**	**1977**	**1990**
Year to	**1962**	**1962**	**1992**	**1992**
Bore (mm)	70	84	70	84
Stroke (mm)	90	90	90	90
Capacity (cc)	346	499	346	499
Inlet valve dia. (in.)	1·50	1·75		
Exhaust valve dia. (in.)	1·38	1·69		
Compression ratio (to 1)	6·5 **1**	6·2 **2**	6·5 **22**	6.5
Valve position	ohv	ohv	ohv	ohv
inlet opens BTDC	30	40		
inlet closes ABDC	60	70		
exhaust opens BBDC	75	75		
exhaust closes ATDC	35	35		
Valve clearance (cold) inlet (in.)	nil	0·001		
Valve clearance (cold) exhaust (in.)	0·002	nil		
Ignition timing (in.)	0·437–0·50	0·312–0·325		
Points gap (in.)	0·012	0·012		
Primary drive chain	$\frac{3}{8}$ duplex	$\frac{3}{8}$ duplex	$\frac{3}{8}$ duplex	3/8 duplex
Rear chain	$\frac{5}{8} \times \frac{3}{8}$	$\frac{5}{8} \times \frac{3}{8}$	$\frac{5}{8} \times \frac{3}{8}$	5/8 x 3/8
O/A ratio: top	5·67 **3**	4·9	5·32	5.00
O/A ratio: 3rd	7·37 **3**	6·37	7·26	6.83
O/A ratio: 2nd	10·2 **3**	8·82	9·80	9.22
O/A ratio: 1st	15·8 **3**	13·65	14·8	13.9
Front tyre (in.)	3·25 × 19 **4**	3·25 × 19	3·25 × 19	3.25 x 19
Rear tyre (in.)	3·25 × 19 **4**	3·50 × 19 **5**	3·25 × 19 **23**	3.50 x 19
Rim front	WM2	WM2		
Rim rear	WM2	WM2		

Model: **Bullet**	**G2**	**500**	**India**	**500 India**
Year from	**1948**	**1952**	**1977**	**1990**
Year to	**1962**	**1962**	**1992**	**1992**
Brake front dia. (in.)	6 **6**	6 **6 7**	6 **24**	7 or 11 disc
Brake front width (in.)	1 **8**	1	1	
Brake rear dia. (in.)	6 **9**	7	6	6
Brake rear width (in.)	1	1	1	
Front suspension	teles	teles	teles	teles
Front movement (in.)	6	6		
Rear type	s/a	s/a	s/a	s/a
Rear movement (in.)	2 **10**	2 **10**		
Petrol tank (Imp. gal.)	3·25 **11**	3·25 **12**	3·2	3.2
Oil tank (Imp. pint)	4·0	4·0	4·0	
Ignition system	magneto **13**	magneto **14**	coil	coil
Generator type	dynamo **15**	dynamo **15**	alternator	alternator
Output (watts)	35 **16**	60	60	
Battery (volt)	6	6	6 **25**	12
Wheelbase (in.)	54	54	54·3	54
Ground clearance (in.)	6·25	6·25	5·5	5.5
Seat height (in.)	29·5	29·5 **17**	30	30
Width (bars) (in.)	27·5		29 **26**	27.6
Length (in.)	84·5		88 **27**	83
Dry weight (lb)	350	370	359	370
Wet weight (lb)		420		
Power: bhp	18 **18**	25 **19**	18 **20**	22
@ rpm	5750 **21**	5250 **19**	5625	5400

1 1955–7·25, 1959–7·75 **2** 1953–6·5, 1959–7·3 **3** 1959–5·2, 7·0, 9·5, 14·3 **4** 1959–3·25 × 17
5 1958–3·25 × 19 **6** 1955–dual, 1959–7 **7** 1960–6 dual **8** 1959–1·5 **9** 1959–7 **10** 1954–3
11 1959–3·75 **12** 1960–4·25 **13** 1960–coil **14** 1959–coil **15** 1956–alternator **16** 1950–60 **17** 1960–31
18 1955–19·5, 1959–20, 1960–21 **19** 1959–27/5750 **20** 1981–17 **21** 1960–6500 **22** 7.25 by 1991
23 Superstar-3.50 x 19 **24** 1988-7 **25** 1987-12 **26** 1987-27.6 **27** 1987-83

Model	500 Twin	Meteor Minor	700 Meteor	Super Meteor
Year from	**1949**	**1958**	**1952**	**1955**
Year to	**1958**	**1963**	**1955**	**1962**
Bore (mm)	64	70	70	70
Stroke (mm)	77	64·5	90	90
Capacity (cc)	495	496	693	693
Inlet valve dia. (in.)	1·375	1·562	1·500	1·562
Exhaust valve dia. (in.)	1·25	1·375		
Comp. ratio (to 1)	6·5 **1**	8·0	6·5 **2**	7·25
Valve position	ohv	ohv	ohv	ohv
inlet opens BTDC	30	30 **3**	30	30
inlet closes ABDC	60	60	60	60
exhaust opens BBDC	75	75	75	75
exhaust closes ATDC	35	35	35	35
Valve clearance (cold) inlet (in.)	0·002	nil	nil	nil
Valve clearance (cold) exhausts (in.)	0·004	nil	nil	nil
Ignition timing (in.)	TDC retard	0·015 retard	0·375–0·437	0·015 retard
Points gap (in)	0·012	0·012	0·012	0·012
Primary drive chain	$\frac{3}{8}$ duplex	$\frac{3}{8}$ duplex	$\frac{3}{8}$ duplex	$\frac{3}{8}$ duplex
Rear chain	$\frac{5}{8} \times \frac{3}{8}$	$\frac{5}{8} \times \frac{3}{8}$	$\frac{5}{8} \times \frac{3}{8}$	$\frac{5}{8} \times \frac{3}{8}$
Sprockets: engine (T)	25		32	20
O/A ratio: top	5·0 **4**	4·67	4·47	4·33 **5**
O/A ratio: 3rd	6·5 **4**	6·08	5·81	5·63 **5**
O/A ratio: 2nd	9·0 **4**	8·42	8·05	7·87 **5**
O/A ratio: 1st	13·9 **4**	13·0	12·45	12·05 **5**
Front tyre (in.)	3·25 × 19	3·25 × 17	3·25 × 19	3·25 × 19
Rear tyre (in.)	3·50 × 19	3·25 × 17 **6**	3·50 × 19	3·50 × 19
Rim front				WM2
Rim rear				WM2
Brake front dia. (in.)	6 **7**	6 **8** **9**	dual 6	dual 6
Brake front width (in.)	1	1 **9**	1 × 2	1 × 2
Brake rear dia. (in.)	6 **10**	7	7	7
Brake rear width (in.)	1	1	1	1
Front suspension	teles	teles	teles	teles
Front movement (in.)	6	6	6	6
Rear type	s/a	s/a	s/a	s/a
Rear movement (in.)	2 **11**	3	2 **11**	3
Petrol tank (Imp. gal.)	3·25	3·75	4·0	4·0 **12**
Oil tank (Imp. pint)	4·0	4·0	4·0	4·0
Ignition system	coil **13** **14**	coil	coil **15**	magneto **16**
Generator type	dynamo **13**	alternator	dynamo	alternator
Output (watts)	60	60	75	70
Battery (volt)	6	6	6	6
Wheelbase (in.)	54	53·5	54	54
Ground clearance (in.)	5·5	6	5·5	5·5
Seat height (in.)	29·5	29·5	29·5	31·0
Width (bars) (in.)	28		28	
Length (in.)	84		83	

Model	500 Twin	Meteor Minor	700 Meteor	Super Meteor
Year from	**1949**	**1958**	**1952**	**1955**
Year to	**1958**	**1963**	**1955**	**1962**
Dry weight (lb)	390 **17**	370	405	410
Wet weight (lb)	410	413		
Power: bhp	25 **18**	30 **19**	36 **20**	40
@ rpm	5500 **18**	6250 **19**	6000	5500

1 1956–7·5 **2** 1955–7·25 not Sports **4** 1957–5·2, 6·7, 9·4, 14·3 **5** 1961–4·44, 5.77, 7·99, 12·35 **6** 1959–3·50
7 1955–dual **8** 1962 only–dual 6 **9** de-luxe–7 × 1·5 **10** 1956–7 × 1 **11** 1954–3 **12** 1961–4·25
13 1957–coil/alternator **14** 1954/5–magneto **15** 1955–magneto **16** 1958–coil **17** from 1951
18 1954–26/6000, 1956–27/6000 **19** Sports–33/6500 **20** 1955–39

Model	Constellation	Interceptor	Interceptor Series II	Rickman
Year from	**1958**	**1962**	**1968**	**1970**
Year to	**1963**	**1968**	**1970**	**1972**
Bore (mm)	70	71 **1**	71	71
Stroke (mm)	90	93	93	93
Capacity (cc)	693	736	736	736
Inlet valve dia. (in.)	1·562			
Exhaust valve dia. (in.)	1·375			
Compression ratio (to 1)	8·5 **2**	8·0	8·5	8·5
Valve position	ohv	ohv	ohv	ohv
Valve timing:		**3**		
inlet opens BTDC	24	35		
inlet closes ABDC	73	82		
exhaust opens BBDC	83	82		
exhaust closes ATDC	35	35		
Valve clearance (cold) inlet (in.)	nil	nil		0·006
Valve clearance (cold) exhaust (in.)	nil	nil		0·007
Ignition timing (in.)	0·375 **4**	0·344		0·344
Points gap (in.)	0·012	0·015		0·015
Primary drive chain	$\frac{3}{8}$ duplex	$\frac{3}{8}$ duplex	$\frac{3}{8}$ duplex	$\frac{3}{8}$ duplex
Rear chain	$\frac{5}{8} \times \frac{3}{8}$	$\frac{5}{8} \times \frac{3}{8}$	$\frac{5}{8} \times \frac{3}{8}$	$\frac{5}{8} \times \frac{3}{8}$
Sprockets: engine (T)	29	29	29	
Sprockets: clutch (T)	56	56	56	
Sprockets: gearbox (T)	20	21	20	
Sprockets: rear (T)	46	46	46	
O/A ratio: top	4·44 **5**	4·22	4·44	4·44
O/A ratio: 3rd	5·77 **5**	5·72	6·05	6·05
O/A ratio: 2nd	7·99 **5**	7·80	8·19	8·19
O/A ratio: 1st	12·35 **5**	11·75	12·4	12·4
Front tyre (in.)	3·25 × 19 **6**	3·25 × 19	3·50 × 19	4·10 × 18
Rear tyre (in.)	3·50 × 19	3·50 × 19	4·00 × 18	4·10 × 18
Rim front		WM2	WM2	WM2 alloy

Model	Constellation	Interceptor	Interceptor Series II	Rickman
Year from	**1958**	**1962**	**1968**	**1970**
Year to	**1963**	**1968**	**1970**	**1972**
Rim rear		WM2	WM3	WM3 alloy.
Brake front dia. (in.)	dual 6	dual 6	8	10 disc
Brake front width (in.	1 × 2	1 × 2		
Brake rear dia. (in.)	7	7	7	9 disc
Brake rear width (in.)	1	1	1	
Front suspension	teles	teles	teles	teles
Front movement (in.)	6	6		
Rear type	s/a	s/a	s/a	s/a
Rear movement (in.)	3	3	3	
Petrol tank (Imp. gal.)	4·25	4·25 **7**	2·0 **8**	3·0
Oil tank (Imp. pint)	4·0	4·0	4·5	4·5
Ignition system	magneto **9**	magneto **10**	cap. + coil	cap. + coil
Generator type	alternator	alternator	alternator	alternator
Output (watts)				120
Battery (volt)	6	6 **11**	12	12
Wheelbase (in.)	54	54 **12**	57	57
Ground clearance (in.)	5·5	5·5	6	6
Seat height (in.)	31	31		
Dry weight (lb)	403	410	426	
Wet weight (lb)	427	420		375
Power: bhp	51 **13**	52·5	52·5	56
@ rpm	6250 **13**	6000	6500	6750

1 actual 70·92 **2** 1961–8·0, 1963–7·0 **3** Sports–37/60/77/35 Supersports–22/75/77/35 **4** 1960–0·437
5 1963–4·93, 6·73, 9·10, 13·77 **6** 1963–3·50 × 19 **7** 1968–2 **8** 1969–4 option **9** 1963–coil **10** 1968–coil
11 1964–12 for de-luxe **12** 1965–57 **13** 1963–40/5500

Model	**250 Clipper**	**350 Clipper**	**350 Clipper**	**350 New Bullet**
Year from	**1953**	**1955**	**1958**	**1963**
Year to	**1957 1**	**1957**	**1962**	**1965**
Bore (mm)	64	70	70	70
Stroke (mm)	77	90	90	90
Capacity (cc)	248	346	346	346
Compression ratio (to 1)	6·5	6·5	6·75	7·5
Valve position	ohv	ohv	ohv	ohv
inlet opens BTDC			30	
inlet closes ABDC			60	
exhaust opens BBDC			75	
exhaust closes ATDC			35	
Valve clearance (cold) inlet (in.)	0·004		nil	
Valve clearance (cold) exhaust (in.)	0·006		nil	
Ignition timing (in.)			0·312	
Points gap (in.)			0·015	
Primary drive chain	$\frac{1}{2}$ **2**		$\frac{3}{8}$ duplex.	
Rear chain	$\frac{1}{2}$ **3**		$\frac{5}{8} \times \frac{3}{8}$	
O/A ratio: top	6·25	5·6	5·7 **5**	5·45
O/A ratio: 3rd	7·37 **4**	7·3	7·4 **5**	6·57
O/A ratio: 2nd	10·2 **4**	10	10·2 **5**	9·27
O/A ratio: 1st	15·8 **4**	15·6	15·8 **5**	15·0
Front tyre (in.)	3·00 × 19	3·25 × 18	3·25 × 19 **7**	3·25 × 17
Rear tyre (in.)	3·00 × 19	3·25 × 18	3·25 × 19 **7**	3·25 × 17
Rim front	WM1			
Rim rear	WM1			
Brake front dia. (in.)	6	6	6	
Brake front width (in.)	1	1	1	
Brake rear dia. (in.)	6	6	6	
Brake rear width (in.)	1	1	1	
Front suspension	teles	teles	teles	teles
Rear type	s/a **8**	s/a	s/a	s/a
Petrol tank (Imp. gal.)	3·25 **9**	3·25	3·25 **10**	3·75 **11**
Oil tank (Imp. pint)	4		4	3
Ignition system	coil	magneto	coil	coil
Generator type	alternator	dynamo	alternator	alternator
Output (watts)	60	60	60	
Battery (volt)	6	6	6	6
Wheelbase (in.)	54	54	54	
Ground clearance (in.)	6	6	5·5	
Seat height (in.)	29·5	29·5	31	
Width (bars) (in.)	28			
Dry weight (lb)	330	362	362	310
Power: bhp	11	15	16·5	22
@ rpm	5500	5500	5500	6500

1 S—1954 **2** S—$\frac{3}{8}$ **3** S—$\frac{5}{8}$ **4** 1955—8·45, 13·1, 18·5 **5** 1960—5·15, 7·03, 9·5, 14·32 **6** S—1954
7 1960—3·25 × 17 **8** S—rigid **9** S—2·75 **10** 1960—3·75 **11** 1964—3·5

Model	**Crusader**	**Clipper II**	**Sports Crusader**	**Crusader Super 5**
Year from	**1956**	**1958**	**1959**	**1961**
Year to	**1962**	**1965**	**1966**	**1963**
Bore (mm)	70	70	70	70
Stroke (mm)	64·5	64·5	64·5	64·5
Capacity (cc)	248	248	248	248
Inlet valve dia. (in.)	1·437		1·562	1·562
Exhaust valve dia. (in.)	1·375		1·375	1·375
Compression ratio (to 1)	7·3 **1**	8·0 **2 3**	8·5 **2**	9·75 **2**
Valve position	ohv	ohv	ohv	ohv
inlet opens BTDC	60	60	50	
inlet closes ABDC	65	65	75	
exhaust opens BBDC	95	95	75	
exhaust closes ATDC	35	35	50	
Valve clearance (cold) inlet (in.)	0·004	nil	0·002	nil
Valve clearance (cold) exhaust (in.)	0·006	0·002	0·004	0·002
Ignition timing (in.)	0·015 retard	0·015 retard	0·218	
Points gap (in.)	0·015	0·015	0·015	
Primary drive chain	$\frac{3}{8} \times \frac{7}{32}$	$\frac{3}{8} \times \frac{7}{32}$	$\frac{3}{8} \times \frac{7}{32}$	$\frac{3}{8} \times \frac{7}{32}$
Rear chain	$\frac{1}{2} \times \frac{5}{16}$	$\frac{1}{2} \times \frac{5}{16}$	$\frac{1}{2} \times \frac{5}{16}$	$\frac{1}{2} \times \frac{5}{16}$
Sprockets: engine (T)	23	23	23	23
Sprockets: clutch (T)	49	49		
O/A ratio: top	5·8 **4**	5·8 **5**	5·8 **5**	5th—6·0
O/A ratio: 3rd	7·8 **4**	7·8 **5**	7·8 **5**	4th—7·5
O/A ratio: 2nd	10·4 **4**	10·4 **5**	10·4 **5**	3rd—9·7
O/A ratio: 1st	17·0 **4**	17·0 **5**	17·0 **5**	2nd—12·8, 1st—17·4
Front tyre (in.)	3·25 × 17	3·25 × 17	3·25 × 17	3·25 × 17
Rear tyre (in.)	3·25 × 17	3·25 × 17	3·25 × 17	3·25 × 17
Rim front	WM2	WM2	WM2	WM2
Rim rear	WM2	WM2	WM2	WM2
Brake front dia. (in.)	6	6	7	7
Brake front width (in.)	1	1	1·5	1·5
Brake rear dia. (in.)	6	6	6	6
Brake rear width (in.)	1	1	1	1
Front suspension	teles	teles	teles	lead. link
Rear type	s/a	s/a	s/a	s/a
Petrol tank (Imp. gal.)	3·0 **6**	3·0 **6 7**	3·75 **7**	3·75
Oil tank (Imp. pint)	3·0	3·0	3·0	3·0
Ignition system	coil	coil	coil	coil
Generator type	alternator	alternator	alternator	alternator
Output (watts)	60	60	60	60
Battery (volt)	6	6	6	6
Wheelbase (in.)	52	52	52	52

Model	Crusader	Clipper II	Sports Crusader	Crusader Super 5
Year from	**1956**	**1958**	**1959**	**1961**
Year to	**1962**	**1965**	**1966**	**1963**
Ground clearance (in.)	5·5	5·5	5·5	5·5
Seat height (in.)	29	29	29	29
Dry weight (lb)	312	312	305	308
Wet weight (lb)				315
Power: bhp	13 **8**	14 **9**	17	20
@ rpm	5750	5750	6250	7500

1 1958–8·0 **2** 1963–9·0 **3** 1959–7·5 **4** 1960–6.1, 8·3, 11·1, 18 **5** 1960–as std. Crusader **6** 1961–3·75 **7** 1964–3·5 **8** 1958–14 **9** 1959–13

Model	250 Trials	Continental	Continental GT	Olympic
Year from	**1961**	**1962**	**1964**	**1964**
Year to	**1963**	**1965**	**1967**	**1965**
Bore (mm)	70	70	70	70
Stroke (mm)	64·5	64·5	64·5	64·5
Capacity (cc)	248	248	248	248
Compression ratio (to 1)	8·75	9·0	9·5	9·0
Valve position	ohv	ohv	ohv	ohv
Valve clearance (cold) inlet (in.)	nil	nil	nil	nil
Valve clearance (cold) exhaust (in.)	0·002	0·002	0·002	0·002
Primary drive chain	$\frac{3}{8} \times \frac{7}{32}$	$\frac{3}{8} \times \frac{7}{32}$	$\frac{3}{8} \times \frac{7}{32}$	$\frac{3}{8} \times \frac{7}{32}$
Rear chain	$\frac{1}{2} \times \frac{5}{16}$	$\frac{1}{2} \times \frac{5}{16}$	$\frac{1}{2} \times \frac{5}{16}$	$\frac{1}{2} \times \frac{5}{16}$
Sprockets: engine (T)	23 **1**			
O/A ratio: top	7·8 **2**	6·02	6·02	6·14
O/A ratio: 3rd	10·6 **2**	7·52	7·52	7·8
O/A ratio: 2nd	19·9 **2**	9·57	9·57	11·05
O/A ratio: 1st	27·2 **2**	2nd—12·82, 1st—17.4	2nd—12·82, 1st—17.4	18.0
Front tyre (in.)	2·75 × 21	3·25 × 17	3·00 × 18	3·25 × 17
Rear tyre (in.)	4·00 × 18	3·25 × 17	3·25 × 17	3·25 × 17
Brake front dia. (in.)		7	7	
Brake front width (in.)		1·5	1·5	
Brake rear dia. (in.)			6	
Front suspension	teles	teles	teles	lead. link
Rear type	s/a	s/a	s/a	s/a
Petrol tank (Imp. gal.)	2·5	3·25 **3**	3·5	3·5
Oil tank (Imp. pint)	3·0	3·0	3·0	3·0
Ignition system	coil	coil	coil	coil
Generator type	alternator	alternator	alternator	alternator
Output (watts)	60	60	60	60
Battery (volt)	6	6	6	6

Model	250 Trials	Continental	Continental GT	Olympic
Year from	**1961**	**1962**	**1964**	**1964**
Year to	**1963**	**1965**	**1967**	**1965**
Wheelbase (in.)	52	52	52	52
Ground clearance (in.)	7·5	5·5	5·5	5·5
Seat height (in.)	29	29·5	29·5	29·5
Dry weight (lb)	285	305		300
Wet weight (lb)			300	
Power: bhp	17	20	21·5	17
@ rpm	6250	7500	7500	6250

1 1963—option 20 **2** 1963—8·0, 13·7, 20·3, 27·8 **3** 1964—3·5

Model	Turbo-Twin	RE	RE2	Ensign
Year from	**1964**	**1945**	**1951**	**1953**
Year to	**1966**	**1950**	**1953**	**1955**
Bore (mm)	50	54	54	56
Stroke (mm)	63·5	55	55	60
Capacity (cc)	249	126	126	148
Compression ratio (to 1)	8·75	5·75 **1**	5·5	6·5
Valve position	two stroke	two stroke	two stroke	two stroke
Ignition timing (in.)	0·155	0·170	0·187	0·170
Points gap (in.)	0·012	0·012	0·015	0·015
Primary drive chain	$\frac{3}{8} \times \frac{7}{32}$	$\frac{3}{8} \times \frac{7}{32}$	$\frac{3}{8} \times \frac{7}{32}$	$\frac{3}{8} \times \frac{7}{32}$
Rear chain	$\frac{1}{2} \times \frac{5}{16}$	$\frac{1}{2} \times \frac{3}{16}$	$\frac{1}{2} \times \frac{3}{16}$	$\frac{1}{2} \times \frac{3}{16}$
Sprockets: engine (T)	20	14	14	14
Sprockets: clutch (T)	43	44	29	29
Sprockets: gearbox (T)		14	14	14
Sprockets: rear (T)		35 **2**	51	47
O/A ratio: top	5·85	7·86 **3**	7·55	6·95
O/A ratio: 3rd	7·75			
O/A ratio: 2nd	11·12	12·77 **3**	12·26	11·30 **4**
O/A ratio: 1st	17·9	22·98 **3**	22·07	20·34 **4**
Front tyre (in.)	3·25 × 17	2·50 × 19 **5**	2·75 × 19	2·75 × 19
Rear tyre (in.)	3·25 × 17	2·50 × 19 **5**	2·75 × 19	2·75 × 19
Rim front				WMO
Rim rear				WMO
Brake front dia. (in.)	6	4	5	5
Brake front width (in.)		0·5	0·62	0·62
Brake rear dia. (in.)	6	5	5	5
Brake rear width (in.)		0·62	0·62	0·62
Front suspension	teles	girder **6**	teles	teles
Rear type	s/a	rigid	rigid	s/a
Petrol tank (Imp. gal.)	3·5	1·5 **7**	1·75	2·0
Oil tank (Imp. pint)	petroil	petroil	petroil	petroil

Model	**Turbo Twin**	**RE**	**RE2**	**Ensign**
Year from	**1964**	**1945**	**1951**	**1953**
Year to	**1966**	**1950**	**1953**	**1955**
Box capacity (Imp. pint)	1·25		0·25	0·25
Chaincase (Imp. pint)			0·125	0·125
Ignition system	fly. mag.	fly. mag.	fly. mag.	fly. mag.
Generator type	alternator	alternator	alternator	alternator
Output (watts)		27	40	40
Battery (volt)	6	3	3	3
Wheelbase (in.)	52	49	48	48
Ground clearance (in.)	6	6·75	5·75	6
Seat height (in.)	29	26·5	28	28
Width (bars) (in.)		26	25·5	
Length (in.)		75		
Dry weight (lb)	298	130 **8**	140	155
Wet weight (lb)	305			
Power: bhp	17	3·5	4·5	5·25
@ rpm	6000	4500	4500	4500

1 1948–5·5 **2** 1948–34 **3** 1948—7·63, 12·40, 22·33 **4** 1955–12, 21·7 **5** 1950—2·75 × 19 **6** 1950—teles **7** 1950—1·75 **8** 1948—135

Model	**Ensign II & III**	**Prince**	**GP5**	**Scrambler**
Year from	**1956 1**	**1958**	**1964**	**1964**
Year to	**1959 2**	**1962**	**1966**	**1965**
Bore (mm)	56	56	66	68
Stroke (mm)	60	60	72	68
Capacity (cc)	148	148	246	247
Compression ratio (to 1)	6·5	6·5	12·5	12·0
Valve position	two stroke	two stroke	two stroke	two stroke
Ignition timing (in.)	0·170	0·187	0·114	
Points gap (in.)	0·015	0·015	0·014	
Primary drive chain	$\frac{3}{8} \times \frac{7}{32}$	$\frac{3}{8} \times \frac{7}{32}$	duplex	
Rear chain	$\frac{1}{2} \times \frac{3}{16}$	$\frac{1}{2} \times \frac{3}{16}$		
Sprockets: engine (T)	14	14		
Sprockets: clutch (T)	29	29		
Sprockets: gearbox (T)	14	14		
Sprockets: rear (T)	47	47		70
O/A ratio: top	6·95	6·95		10
O/A ratio: 3rd				12·6
O/A ratio: 2nd	12·0	11·75		16·7
O/A ratio: 1st	21·7	20·35		25·4
Front tyre (in.)	2·75 × 19	2·75 × 19	2·75 × 18	2·75 × 21
Rear tyre (in.)	2·75 × 19	2·75 × 19	3·25 × 18	4·00 × 18

Model	Ensign II & III	Prince	GP5	Scrambler
Year from	**1956 1**	**1958**	**1964**	**1964**
Year to	**1959 2**	**1962**	**1966**	**1965**
Rim front	WM0		alloy	
Rim rear	WM0		alloy	
Brake front dia. (in.)	5	5		
Brake front width (in.)	1	1		
Brake rear dia (in.)	5	5		
Brake rear width (in.)	1	1		
Front suspension	teles	teles	lead. link	lead. link
Rear type	s/a	s/a	s/a	s/a
Petrol tank (Imp. gal.)	2·0	3·0	7·0	2·5
Oil tank (Imp. pint)	petroil	petroil	petroil	petroil
Box capacity (Imp. pint)	0·25			
Chaincase (Imp. pint)	0·125			
Ignition system	fly. mag.	fly. mag.	coil	fly. mag.
Generator type	alternator	alternator		
Output (watts)	40			
Battery (volt)	3 **3**	6	6	
Wheelbase (in.)	48	48		
Ground clearance (in.)	6	6		
Seat height (in.)	28·5	28·5		
Dry weight (lb)	155 **4**	200		
Wet weight (lb)	183			
Power: bhp	6	7·5	34	22
@ rpm	4750	4750	8000	

1 III—1958 **2** II—1958 **3** 1957—6 **4** III—168

Colours

1945/46/47

RE

Black frame, forks, rims, hubs, mudguards, carrier, toolbox, headlamp shell, outer chaincase, chainguard; tank black with silver lines and motif, round RE badge on left side only; chrome plated exhaust pipe, silencer, controls, handlebars and headlamp rim.

G & J

All black except tank chrome plated with frosted silver panels lined in red, Royal Enfield on panels in bright red; chrome plated headlamp rim, exhaust system, handlebars and controls except rear brake pedal; rims in black, front brake back plate polished alloy.

1948

RE

Chrome plated wheel rims, saddle springs and points cover; in mid-year colour changed to maroon and silver.

G & J

Chrome plated fork legs and wheel rims, rim centres in black.

350 Bullet prototype

Polychromatic green frame and rear fork, tank chrome plated with green panels carrying Enfield name, alloy mudguards and stays, chrome plated wheel rims.

1949

RE

As 1948. Alternatives silver grey, battleship grey or maroon.

G & J

Polished fork ends.

350 Bullet

Silver grey frame, forks, mudguards, toolbox and hubs; chrome plated tank with frosted silver panels, lined and with name in red; chrome plated wheel rims; black headlamp shell and air cleaner. Chrome plated mudguards on trials model.

500 Twin

As **Bullet**, tank lined in red, wheel rims with silver grey centres lined red.

1950

RE

As 1949 but chrome plated tank panels.

G, **J**, **350 Bullet** and **500 Twin**

As 1949.

1951

RE2

As 1950 **RE** except black headlight shell, tank with silver panels, 'Royal Enfield' in red.

G & J

Black tank with gold lining, chrome plated tank top strip.

350 Bullet

As 1949, during year black replaced silver grey.

500 Twin

As **Bullet**.

1952

RE2

Wheel rims grey with silver lining, rest as in 1951.

G, **J** and **350 Bullet**
As 1951.

500 Twin
Finish as 1949 but in polychromatic grey and frame in black.

1953
RE2
Blue grey, tank transfer with Royal Enfield set in 5 lines, rest as 1952.

Ensign
Polychromatic copper beech with matching saddle top, black headlight, bright zinc plated rims, tank transfer as **RE2**.

Clipper
Olive green frame, forks, tank, toolboxes and mudguards, tank with pressed metal wing badge, chrome plated rims.

S
Black and chrome finish.

G, **J** and **350 Bullet**
Pressed metal wing tank badges.

500 Bullet
Polychromatic copper beech frame, forks, tank, mudguards, toolboxes, rear unit tops; black air cleaner and headlamp shell; chrome plated headlamp rim and lower rear units; bright zinc plated wheel rims; tank badge as **350**.

500 Twin
Polychromatic blue grey tank, mudguards, toolboxes, forks; black frame, headlamp shell and air cleaner; pressed metal wing tank badges.

700 Meteor
Polychromatic copper beech for all painted items with matching dualseat, chrome plated rims, pressed metal wing tank badges.

1954
Ensign
AS 1953 except wheel rims chrome plated.

Clipper, **S**, **G**, **J**, **500 Bullet** and **Meteor**
As 1953. **Meteor** with Burgundy/black alternative.

350 Bullet and **500 Twin**
In silver grey.

1955
Ensign
All painted parts in maroon, options of olive green or polychromatic silver grey; chrome plated rims.

Clipper
As **Ensign**, olive green standard, tank badges as 1953.

J
As 1953.

Bullet
As **Ensign**, **350** in silver grey and **500** in maroon as standard, tank badges as 1953.

Twins
As **Ensign**, **500** in silver grey and **700** in maroon as standard, tank badges as 1953. **Meteor** has olive green option.

1956
Ensign II
Surf green, chrome plated rear spring covers, plastic tank badges.

Clipper
As 1955 in maroon.

350 Clipper
As **250** in olive green.

Crusader
Surf green frame, forks, mudguards, toolboxes and tank which had chrome plated panels and plastic badges, chrome plated rims.

Bullet
Maroon, plastic tank badges, chrome tank panels introduced although not shown in programme.

500 Twin and **Super Meteor**
As **Bullets**.

1957
Ensign II, **Clippers**, **Bullets**, **500 Twin** and **Super Meteor**.
As 1956.

Crusader
Black, green or maroon options.

1958
Ensign II & **III**
Surf green as 1956 with black as option.

Crusader
Black standard with chrome plated rims and tank panels. Two-tone options with tank, mudguards, toolbox lids and chainguard in polychromatic burgundy, Wedgewood blue or silver grey, and frame, forks and toolbox in black.

Clipper II
Black with chrome plated rims, tank gold lined with plastic badges.

350 Clipper
Black with chrome plated rims and wing tank badges.

Bullets, **500 Twin** and **Super Meteor**
As **Crusader**.

Meteor Minor
De-luxe as **Crusader**, standard in black only.

Constellation
Tank chrome plated with polychromatic burgundy top panel and matching toolbox lids, plastic tank badges; chrome plated mudguards and rims; black frame, forks and toolbox.

1959
Ensign III
Surf green, black frame and forks.

Prince
Cherry red tank, mudguards, side covers and chaincase; black frame and forks; gold lining. Bottom half of tank black. Alternative of mist grey and black.

Crusader and **Crusader Sports**
As 1958 in polychromatic burgundy or peacock blue.

Clipper II and **350 Clipper**
As **Crusader** in cherry red.

Bullets
As **Crusader** except for chrome plated mudguards.

Meteor Minor
As 1958 but with de-luxe in 1959 **Crusader** colours and standard in cherry red.

Super Meteor and **Constellation**
In burgundy or peacock blue.

1960
Prince, **Crusader**, **Crusader Sports**, **Clipper II**, **Bullets**, **Meteor Minor de-luxe**, **Super Meteor** and **Constellation**
As 1959.

350 Clipper
As **Crusader** with two-tone finish of mist grey for mudguards and tank top, black for tank bottom and other items.

Meteor Minor Sports
As **de-luxe** with chrome plated tank with top panel painted in de-luxe colours, plastic tank badges, chrome plated mudguards, black frame, forks, toolbox.

1961
Prince, **350 Clipper**, **Crusader**, **Bullets** and **Meteor Minor de-luxe**
As 1960.

Clipper II and **Meteor Minor Sports**
As **350 Clipper**.

Crusader Sports
As **Crusader** except tank chrome plated with painted top panel in 1959 colours, chrome plated mudguards.

Super Meteor
Tank as **Constellation** in 1959 colours. Rear enclosure in machine colour. Tank painted, not plated (like Connie).

Constellation
As 1959, rear enclosure in machine colour.

1962
Prince
Two-tone with burgundy tank top and mudguards, cream tank bottom and side covers, black frame and forks.

Crusader, **Clipper II**, **350 Clipper**, **Meteor Minor de-luxe** and **Super Meteor**
As **Prince**, **Super Meteor** with rear enclosure in burgundy.

Crusader Sports
As 1961 except rear mudguard painted and front chrome plated.

Super 5
Marina grey frame, forks, casquette, mudguards, toolbox and lids, tank top panel; red flashes and lines on mudguards and forks; tank chrome plated with panel red lined; rims chrome plated; seat with black sides and fine chequered grey top; black fork link covers. These were later changed to grey and the frame to black.

250 Trials
Black frame, forks, toolbox and lids; alloy hubs and mudguards.

Bullets
As 1959.

Meteor Minor Sports and **Constellation**
As **Crusader Sports** with black rear side pressings.

1963
Clipper II
Two-tone in flame red and cream as in 1962.

Crusader Sports
As 1959 in polychromatic blaze or gold.

New Bullet and **Super 5**
As **Crusader Sports** and in same colours.

250 Trials
As 1962.

Continental
As **Crusader Sports** with checker tape on fork legs between crowns.

500 Sports Twin
Chrome plated tank with gold top panel, gold mudguards and toolbox lids, black frame, forks and toolbox.

Constellation
Two-tone tank in flame and cream with flame mudguards, toolbox lids and rear enclosure; chrome plated rims; black frame, forks and toolbox.

Interceptor
Tank chrome plated with polychromatic blaze top panel; blaze toolbox lids and chainguard; chrome plated mudguards and rims; black frame, forks and toolbox; plastic tank badges.

1964
Clipper II
As 1963.

New Bullet and **Turbo Twin**
As **Clipper**.

Turbo Twin Sports
Chrome plated mudguards and tank panels.

Crusader Sports and **Continental**
As 1963 in flame red or hi-fi blue for mudguards and tank.

Interceptor
De-luxe as 1963 or in polychromatic blue; standard with polychromatic silver mudguards and tank side panels.

1965
Turbo Twin
New tank style in flame red and cream with gold lining, broad cream stripe with flame side panels, flame mudguards, black toolbox lids.

Turbo Twin Sports
As standard with chrome plating as 1964.

Continental
In white and chrome as well as existing blaze or blue.

Crusader Sports
New tank style.

Clipper
Red toolbox lids, cream tank centre.

Bullet
Eggshell blue for tank sides, mudguards and chainguard; white tank stripe and toolbox lids.

Olympic
Flamboyant blaze or blue tank top and rear mudguard; silver front fork, tank bottom and panels; black frame, fork link covers, toolbox and lids.

GT
Silver frame, red mudguards and tank with 'Royal Enfield' in white on sides.

1966
Crusader Sports and **Turbo Twin Sports**
As 1965.

GT
As 1965 except black frame.

1969/70
Interceptor Series II
Chrome plated tank, mudguards, headlamp shell and rim, wheel rims; black frame and forks; tank badge with 'Royal' in red script and 'Enfield' in black block. UK version with large tank, painted red with plastic badges.

1970/71/72
Rickman Metisse and **Enfield Rickman**
Nickel plated frame; chrome plated headlamp shell; tank, seat unit and front mudguard in single colour of orange, blue, red or green. The tool and battery box was in white.

1977
Indian 350
In black with chrome plated side panels on cream lined tank, yellow badges with 'Enfield India' in red on flying wing formed in panel with red infill, chrome plated mudguards and rims. Several imported with red or pale green cycle parts.

1981/82
Indian 350
As 1977 except tank black with gold lining and gold 'Enfield' name on sides.
Late in year mudguards in black.

Numbering system

This series of books normally contains as full a list as possible of engine and frame numbers for the machines described. This has not proved to be practical with Royal Enfield and it is with regret that the author and the publisher have omitted them.

The reasons are two-fold. First the numbering sequence is known to be random to some degree and not to follow any pattern which would enable the information to be set out in an orderly fashion. Second is that, due to this, it would take a great deal of space to provide the data and this would severely overstretch the bounds of the book's pages.

Fortunately the records do exist and are contained in some 37 ledgers, each 27 × 18 in. and 3 in. thick, all held by the Royal Enfield Owners Club dating officer. He can provide the date of dispatch, and often the destination, of any machine on receipt of the engine and frame number. He is aided in this if the owner also gives an indication of the model he has by listing the number of cylinders and basic frame specification.

Carburettor settings

Year	Model	Amal Type	Size	Main	Pilot	Slide	Needle Pos.	Needle Jet
1940–54	125 cc RE	223	$\frac{11}{16}$	85		5	3	·107
1948	125 cc RE	223	$\frac{11}{16}$	90		5	3	·107
1955–56	125 cc RE2	223	$\frac{11}{16}$	90		5	3	·107
1953–54	150 cc Ensign	223	$\frac{3}{4}$	110		5/D	3	·107
1955–61	150 cc Ensign & Prince	223	$\frac{3}{4}$	100		5	3	·107
1957–61	150 cc Madras model	223	$\frac{3}{4}$	90		5	3	·107
1940–45	250 cc sv, D, ex-WD	274	$\frac{23}{32}$	75		4/5	3	Std
1954–55	250 cc Clipper	274	$\frac{25}{32}$	75		4/4	2	Std
1955–61	250 cc Clipper I	375	$\frac{25}{32}$	120	25	4	3	·105
1957–60	250 cc Clipper II	375	$\frac{7}{8}$	120	25	$3\frac{1}{2}$	3	·105
1962	250 cc Clipper Series I	375	$\frac{25}{32}$	120	25	4	3	·105
1962–63	250 cc Clipper Series II	375	$\frac{7}{8}$	120	25	$3\frac{1}{2}$	3	·105
1964–66	250 cc Clipper and Olympic	375	$\frac{7}{8}$	95	25	$3\frac{1}{2}$	3	·105
1955–61	250 cc Crusader	375	$\frac{7}{8}$	120	25	$3\frac{1}{2}$	3	·105
1962	250 cc Crusader	376	$1\frac{1}{16}$	320	25	$3\frac{1}{2}$	3	·106
1958–63	250 cc Crusader Sports	376	$\frac{15}{16}$	150	25	$3\frac{1}{2}$	3	·106
1962–63	250 cc Trials	376	$\frac{15}{16}$	150	25	$3\frac{1}{2}$	3	·106
1964–66	250 cc Crusader Sports	376	$\frac{15}{16}$	130	25	$3\frac{1}{2}$	3	·106
1963	250 cc Continental & Super 5	376	$1\frac{1}{16}$	180	25	$3\frac{1}{2}$	4	·106
1964–65	250 cc Continental	376	$1\frac{1}{16}$	170	25	$3\frac{1}{2}$	4	·106
1965	250 cc Continental GT	376	$1\frac{1}{16}$	180	25	$3\frac{1}{2}$	4	·106
1965–66	250 cc GT	389	$1\frac{1}{8}$	260	30	$3\frac{1}{2}$	3	·106
1965–66	250 cc GP	GP2	$1\frac{1}{2}$	520	25	4	1	·107
1940–45	350 cc sv, C, ex-WD	274	$\frac{25}{32}$	85		4/5	3	Std
1940–45	350 cc ohv, CO, ex-WD	276	$\frac{15}{16}$	130		6/4	2	Std
1946–54	350 cc G	276	$\frac{15}{16}$	130		6/4	3	Std
1946–50	500 cc J	276	$\frac{15}{16}$	140		6/4	3	Std
1948	500 cc J	276	1	150		6/4	3	Std
1947–55	500 cc J2	276	$1\frac{1}{16}$	170		6/4	2	Std
1947–54	350 cc Bullet G2	276	1	140		6/4	3	Std
1952–54	350 cc Bullet G2	TT9	$1\frac{1}{16}$	300		6	4	·109

Year	Model	Amal Type	Size	Main	Pilot	Slide	Needle Pos.	Needle Jet.
1955–57	350 cc Bullet	376	1	180	30	4	3	·106
1955–56	350 cc Bullet	GP	$1\frac{1}{16}$	220		6	3	·109
1956	350 cc Bullet	TT9	$1\frac{1}{16}$	300		6	4	·109
1958–61	350 cc Bullet	376	$1\frac{1}{16}$	170	30	4	3	·106
1960–61	350 cc Bullet	376	1	180	30	4	3	·106
1960–61	350 cc Bullet (USA)	389	$1\frac{3}{16}$	260	30	4	3	·106
1962	350 cc Bullet	376	$1\frac{1}{16}$	170	30	4	3	·106
1963	350 cc Bullet	376	$1\frac{1}{16}$	220	25	$3\frac{1}{2}$	3	·106
1964–66	350 cc Bullet	376	$1\frac{1}{16}$	160	25	$3\frac{1}{2}$	3	·106
1955–57	350 cc Clipper	276	$\frac{15}{16}$	130		6/4	3	·106
1957–61	350 cc Clipper	376	1	180	30	4	3	·106
1953–55	500 cc Bullet	289	$1\frac{1}{8}$	180		29/3	2	Std
1955–60	500 cc Bullet	389	$1\frac{1}{8}$	200	30	$3\frac{1}{2}$	2	·106
1956	500 cc Scrambler	TT9	$1\frac{3}{16}$	360		6	4	·109
1959–62	500 cc Bullet	389	$1\frac{3}{16}$	230	30	$3\frac{1}{2}$	3	·106
1947–48	500 cc Twin	276	$\frac{15}{16}$	140		6/3	2	·109
1949–55	500 cc Twin	276	$\frac{15}{16}$	150		6/4	2	·109
1955–58	500 cc Twin	376	$\frac{15}{16}$	230	25	$3\frac{1}{2}$	2	·106
1957–63	500 cc Meteor Minor & Sports Twin	376	$1\frac{1}{16}$	250	30	$3\frac{1}{2}$	2	·106
1952–55	700 cc Meteor	276	$1\frac{1}{16}$	170		6/4	3	·106
1957–61	700 cc Super Meteor	376	$1\frac{1}{16}$	240	30	$3\frac{1}{2}$	3	·106
1957–60	700 cc Super Meteor	389	$1\frac{1}{8}$	270	30	$3\frac{1}{2}$	2	·106
1957–58	700 cc Constellation	TT9	$1\frac{3}{16}$	480		5	3	·109
1959–62	700 cc Constellation	389	$1\frac{3}{16}$	300	30	$3\frac{1}{2}$	2	·106
1959–61	700 cc Constellation (pair)	376	$1\frac{1}{16}$	320	25	4	3	·106
1960–61	700 cc Constellation (pair)	389	$1\frac{3}{16}$	380	30	$3\frac{1}{2}$	3	·106
1962–63	700 cc Super Meteor & Constellation	376	$1\frac{1}{16}$	240	30	$3\frac{1}{2}$	3	·106
1959–60	700 cc Berkeley Twin	376	$1\frac{1}{16}$	240	30	$3\frac{1}{2}$	3	·106
1959	700 cc Berkeley Twin	TT9	$1\frac{3}{16}$	480		5	3	·109
1963–66	750 cc Interceptor (pair)	389	$1\frac{3}{16}$	380	25	$3\frac{1}{2}$	3	·106
1965–66	750 cc Interceptor	389	$1\frac{3}{16}$	400	30	3	3	·109
1965	750 cc Interceptor (USA police)	389	$1\frac{3}{16}$	360	25	$3\frac{1}{2}$	3	·106
1967–68	750 cc Interceptor (pair)	930	30 mm	220	20	3	2	·107

Prices

The prices of the machines and some of their accessories over the years are set out in the tables below:

Date	G	J	J2	350 Bullet	500 Bullet	350 Trials
Nov. 45	£93 + PT	£103 + PT				
27. 6.46	£118 2*s*. 3*d*.	£130 16*s*. 3*d*.				
4. 7.46	£133 7*s*. 0*d*.	£146 1*s*. 0*d*.				
2.10.47	£146 1*s*. 0*d*.					
4.11.48	£146 1*s*. 0*d*.			£171 9*s*. 0*d*.		
2. 2.50	£152 8*s*. 0*d*.		£171 9*s*. 0*d*.	£177 16*s*. 0*d*.		
8. 3.51	£158 15*s*. 0*d*.		£171 9*s*. 0*d*.	£184 3*s*. 0*d*.		
11.10.51	£166 2*s*. 3*d*.		£185 5*s*. 7*d*.	£191 13*s*. 4*d*.		
2.10.52	£166 2*s*. 3*d*.		£185 5*s*. 7*d*.	£191 13*s*. 4*d*.	£214 0*s*. 7*d*.	
8.10.53	£162 0*s*. 0*d*.		£174 0*s*. 0*d*.	£186 0*s*. 0*d*.	£204 0*s*. 0*d*.	
21.10.54			£174 0*s*. 0*d*.	£192 0*s*. 0*d*.	£204 0*s*. 0*d*.	
22. 9.55				£204 0*s*. 0*d*.	£234 0*s*. 0*d*.	
18.10.56				£210 16*s*. 0*d*.	£223 4*s*. 0*d*.	
3.10.57				£230 15*s*. 9*d*.	£244 10*s*. 2*d*.	
27.11.58				£235 0*s*. 0*d*.	£245 0*s*. 0*d*.	£242 10*s*. 0*d*.
22.10.59				£227 4*s*. 7*d*.	£236 17*s*. 11*d*.	£234 9*s*. 7*d*.
29. 9.60				£239 16*s*. 0*d*.	£251 2*s*. 6*d*.	
21. 9.61				£244 19*s*. 6*d*.	£255 10*s*. 0*d*.	
				Indian		
June 77				£695		
Dec. 78				£710		
Aug. 79				£756		
June 80				£899		
Aug. 81				£968		

Date	500 Twin		Meteor	Super Meteor	Constellation	Interceptor
4.11.48	£196 17s. 0d.					
16.11.50	£212 14s. 6d.					
8. 3.51	£219 1s. 6d.					
11.10.51	£233 3s. 11d.					
2.10.52	£233 3s. 11d.		£245 19s. 5d.			
8.10.53	£222 0s. 0d.		£234 0s. 0d.			
22. 9.55	£216 0s. 0d.			£249 12s. 0d.		
18.10.56	£235 12s. 0d.			£260 8s. 0d.		
3.10.57	£258 17s. 1d.			£283 16s. 1d.		
	Meteor Minor					
	standard	**de-luxe**				
3. 4.58	£249 10s. 0d.	£274 9s. 0d.			£295 0s. 0d.	
16.10.58	£249 10s. 0d.	£270 0s. 0d.		£275 0s. 0d.	£295 0s. 0d.	
	Sports					
22.10.59	£270 0s. 0d.	£261 1s. 6d.		£266 18s. 2d.	£285 4s. 11d.	
29. 9.60	£254 9s. 6d.	£261 14s. 6d.		£280 0s. 0d.	£299 10s. 0d.	
			500 Sports Twin			
21. 9.61	£260 0s. 0d.	£266 10s. 0d.		£287 10s. 0d.	£305 0s. 0d.	
3.10.62			£250 0s. 0d.		£285 0s. 0d.	£310 0s. 0d.
17.10.63						£329 17s. 6d.
22.10.64						£349 10s. 0d.
16. 9.65						£358 10s. 0d.
						Series II
Dec. 69						£455
1971						£625
1972						£750

Date	250 Clipper	350 Clipper	Crusader	Crusader Sports	Super 5	250 Trials	
8.10.53	£162 0s. 0d.						
22. 9.55	£171 0s. 0d.	£183 0s. 0d.					
18.10.56	£161 4s. 0d.	£173 12s. 0d.	£198 8s. 0d.				
		New 350					
3.10.57	**Clipper II**	£199 18s. 3d.	£212 1s. 6d.				
16.10.58	£179 19s. 6d.	£199 18s. 3d.	£205 0s. 0d.	£210 0s. 0d.			
22.10.59	£174 0s. 5d.	£193 6s. 0d.	£198 4s. 5d.	£203 1s. 2d.			
29. 9.60	£189 10s. 0d.	£219 17s. 6d.	£209 10s. 0d.	£219 7s. 6d.			
21. 9.61	£185 0s. 0d.	£224 19s. 6d.	£213 10s. 0d.	£224 19s. 6d.	£239 10s. 0d.	£215 0s. 0d.	
		New Bullet	**Continental**				
3.10.62	£195 0s. 0d.	£239 17s. 6d.	£242 10s. 0d.	£235 0s. 0d.	£245 0s. 0d.	£226 15s. 0d.	**Turbo Twin**
17.10.63	£199 17s. 6d.	£239 19s. 6d.	£247 12s. 6d.	£235 10s. 0d.	**Olympic**	**Turbo Twin Sports**	£195 0s. 0d.
22.10.64	£211 19s. 6d.	£252 10s. 0d.	£261 19s. 6d.	£248 17s. 6d.	£225 0s. 0d.	£215 0s. 0d.	£206 0s. 0d.
7.11.64			**GT** £270				
16. 9.65	£217 10s. 0d.		**GT** £275	£254 17s. 6d.		£219 10s. 0d.	

Date	125 RE	150 Ensign	Ensign II	Ensign III	Prince
Nov. 45	£48 + PT				
27. 6.46	£60 19s. 3d.				
4. 7.46	£66 0s. 9d.				
2.10.47	£73 13s. 3d.				
2. 2.50	£73 17s. 8d.				
16.11.50	£82 11s. 0d.				
8. 3.51	£86 9s. 2d.				
11.10.51	£92 0s. 0d.				
2.10.52	£88 3s. 4d.	£99 10s. 2d.			
8.10.53		£93 12s. 0d.			
22. 9.55		£99 19s. 6d.			
1. 3.56			£105 8s. 0d.		
18.10.56			£ 99 19s. 6d.		
3.10.57			£109 3s. 1d.	£115 17s. 3d.	
27.11.58				£115 17s. 3d.	£121 10s. 0d.
22.10.59				£112 0s. 7d.	£117 9s. 8d.
29. 9.60					£124 7s. 6d.
21. 9.61					£126 10s. 0d.

Date	RE legshields	RE speedo	G & J speedo	Trials Bullet light set	Sidecar spec.	Dualseat	Panniers
3. 1.46	£ 1 5s. 0d.	£ 2 15s. 0d.	£ 3 10s. 0d.				
2.10.47	£ 1 5s. 0d.	£ 4 0s. 8d.	£ 5 1s. 7d.				
16.11.50	£ 1 12s. 6d.			£ 9 10s. 6d.	£3 19s. 11d.		
11.10.51	£ 1 15s. 0d.	**Clipper**	**Clipper**	£10 17s. 3d.	£4 0s. 6d.	£5 2s. 3d.	£ 6 7s. 9d.
3. 3.55	£ 2 5s. 0d.	**dualseat**	**prop-stand**	£10 4s. 0d.	£2 5s. 0d.	**Q.D. rear**	£ 6 12s. 0d.
22. 9.55	£ 2 5s. 0d.	£ 4 10s. 0d.	£1 10s. 0d.		£2 10s. 5d.	£3 0s. 0d.	£ 7 10s. 0d.
3.10.57	£ 3 10s. 0d.	£ 6 4s. 9d.	£1 17s. 5d.		£3 2s. 4d.		£ 9 7s. 1d.

	Airflow on 150	Airflow 250–700	Chain-case	Air cleaner	Prop-stand	Pannier	Sports-flow
16.10.58	£22 10s. 0d.	£25 0s. 0d.	£5 0s. 0d.	£1 1s. 0d.	£1 17s. 5d.	£9 7s. 1d.	
29. 9.60	£24 8s. 3d.	£30 14s. 3d.	£5 2s. 6d.	£1 14s. 0d.	£2 0s. 0d.	£9 9s. 0d.	
21. 9.61	£24 16s. 7d.	£31 4s. 9d.	£5 4s. 3d.	£1 14s. 7d.	£1 4s. 6d.	£9 12s. 2d.	
3.10.62		£30 0s. 0d.	£5 4s. 6d.	£1 15s. 0d.	£1 7s. 0d.	£9 19s. 0d.	£18 18s. 0d.
16.10.63		£33 0s. 0d.	£5 4s. 6d.	£1 17s. 6d.	£1 7s. 6d.		£18 18s. 0d.

Model recognition points

This section has been compiled for use with the data in the other appendices and represents a precis of the main text as applicable.

1945
RE: Single cylinder two stroke, flywheel magneto on right, alloy head, 3 speed handchange, rigid frame, rubber band girders, saddle, carrier, centre stand, tubular toolbox.
G: 350 cc, ohv, iron head and barrel, oil filler cap in front of barrel, rigid frame, teles, sprung front guard, upswept pipe option.
J: 500 cc version of **G**, 3·50 section rear tyre.

1946
No changes.

1947
RE: No changes.
G & J: Optional larger petrol tank, finned collar on exhaust pipe.

1948
RE: Expansion exhaust pipe, gearing raised, angled speedo, longer toolbox.
G & J: Pillion rest standard, gearing lowered.
350 Bullet (prototype): Semi-unit construction, swinging fork, alloy chaincase, oil filler cap behind timing case.

1949
RE: No changes.
G: No ribs on front brake back plate.
J2: Twin port version of **J** with dual silencers, change as **G**.
350 Bullet (road): As 1948 with road equipment. Also available in trials or scrambles forms.
500 Twin: New model. Semi-unit twin in **Bullet**-type frame. Saddle, sprung front mudguard, ammeter and switch in box, separate air cleaner box, distributor on dynamo, separate headlamp, two toolboxes.

1950
RE: Teles fitted to girder mountings, deeper fuel tank, bigger section tyres, screw fastener for toolbox.
G: 60 watt dynamo, single lifting rear stay, one toolbox.
J2: As **G** except two toolboxes, sidecar forks and gearing option.
350 Bullet: Cylindrical silencer, cast alloy fork top cover carrying speedo, no valance on front guard.
500 Twin: Cast alloy fork top and mudguard as Bullet, switch on front face of box.

1951
RE2: New model with new engine unit, very smooth exterior, engine clutch, 3 speed, footchange, cradle frame, teles, saddle, tubular toolbox, external ignition coil, rear carrier, expansion exhaust pipe.
G: New fork ends, unsprung front mudguard with three stays.
J2: As **G** plus new fork yokes for sidecar use and steering damper.
350 Bullet: Fork ends and mudguard as **G**.
500 Twin: Die cast crankcase halves, weight down, fork ends and mudguard as **G**.

1952
RE & G: No changes.
J2: Top fork yoke as **G**, spring prop-stand.
350 Bullet: Breather on chaincase with banjo union, alloy barrel on trials model, prop-stand as **J2**, dualseat option.
500 Twin: External oil feed to rockers, added crankcase screws, ammeter and light switch in headlamp, ignition

switch in toolbox, bigger air cleaner box, breather on chaincase, dualseat option.

1953

RE: No changes, discontinued in year.
Ensign: New model of 150 cc, swinging fork frame with plunger type springs, shaped toolbox, saddle, valanced mudguards, bigger tank.
G & **J2**: Embossed tank motif, floating brake cam spindles.
350 Bullet: Engine internals altered.
500 Bullet: New model, four main bearings, oil system of improved type, breather on crankcase mouth, frame as **350,** 7 in. rear brake, underslung pilot lamp.
500 Twin: Die cast heads, new inlet manifóld, underslung pilot lamp, new stop and tail lamp.
700 Meteor: New model, enlarged **500 Twin**, dual front brake, 7 in. rear, 4 gallon tank, breather on crankcase wall, valanced mudguards, underslung pilot lamp.
250 Clipper: New model as **G** in swinging fork frame, teles, saddle, casquette.
S: New model, **Clipper** engine and gearbox in **G** cycle parts, rigid, no casquette.

1954

Ensign, **250 Clipper**, **G**, **S** & **J2**: No changes, **G** and **S** discontinued late in year.
350 & **500 Bullets**: Casquette, silencer parallel to ground, improved rear units.
500 Twin: Casquette, coil or mag-dyno option.
Meteor: Casquette, longer fork ends for sidecar use, proprietary rear units, front mudguard with tubular stay doubling as front stand.

1955

Ensign: Clutch changes, covers over rear springs. In autumn wider brakes, fork changes; then model replaced.
250 Clipper: Oval air cleaner box, new gearbox end cover with lower pedal position.
J2: No changes, discontinued in year.
350 Bullet: Gearbox with pedals on concentric spindles, dualseat, dual front brake, new air cleaner, racing option.
500 Bullet: As **350**, new cam forms, new frame lugs.
500 Twin: Gearbox as **Bullet**, dual front brake, oval air box, mag-dyno standard, qd rear option in year.
Meteor: Gearbox as **Bullet**, oval air box, mag-dyno, qd rear option in year.

1956

Ensign II: As **Ensign**, more fins on head and barrel, plug at rear of head, decompressor at front, dualseat, coil under seat.
250 Clipper: Monobloc, Lucas electrics, points in housing, ignition switch in toolbox.
350 Clipper: New model. As **250**, casquette with pilots, high gear pedal pivot, neutral finder, mag-dyno, 276 carb, saddle.
350 & **500 Bullets**: New cylinder head, crankcase common, alternator, magneto, new frame, air cleaner on right, tools and battery on left, folding kickstart, boxed in rear number plate, full width alloy qd rear as extra, new tank mounting.
500 Twin: Full width qd rear, new tank mounting, folding kickstart.
700 Super Meteor: New model based on Meteor. Improved power output, alternator, magneto, new frame, new tank mounting, folding kickstart, two start oil pump drive, air filter and tools on left, battery on right.
250 Crusader: New model introduced in August. Unit construction with oil within crankcase, 17 in. wheels, deep mudguards, chaincase, dualseat, all iron engine, centre section with two lids for air cleaner, tools, battery and ignition switch.

1957

Ensign II: Available with battery, rectifier, electric horn, stop light and ammeter as option.
250 & **350 Clippers**, **250 Crusader**: No changes, **350 Clipper** discontinued in year.
Bullets: Larger air cleaner element, qd rear standard.
500 Twin: Monobloc, ac electrics, distributors in magneto position, frame as **1956 Bullet**, special rear mudguard mounting.
Super Meteor: Rear mudguard as **500 Twin**, improved magneto.

1958

Ensign II: No changes, discontinued in year.
Ensign III: New model as **II** with battery option, cast alloy top crown, longer headlight with speedometer, dualseat.
250 Clipper: Discontinued early in year.
250 Clipper II: New model, cheap **Crusader**, no valance on front mudguard, saddle, no polish to castings, no chaincase, chainguard on top run only.
250 Crusader: Minor changes to gearchange and lubrication, Burgess silencer.

350 Clipper: New model with **Bullet** bottom half, iron head and barrel, alternator, coil, swinging fork, **Crusader**-type forks, full width **Crusader** front brake, saddle, old silencer, casquette, Monobloc, 19 in. wheels, dualseat option.

Bullets: Burgess silencer.
500 Twin: New silencers, discontinued early in year.
Meteor Minor: New model, short stroke, alternator, coil, Monobloc, casquette, **Crusader** frame, new clutch mechanism, siamezed pipes, 17 in. wheels. Standard model—6 in. front brake, saddle.
De-luxe—7 in. front, dualseat, enclosed chain, air cleaner, stop light, prop stand, valanced front mudguard with numbers on sides, pressing over points unit. From July Airflow available as option.
Super Meteor: New silencers, no tailpipes, coil, distributor.
Constellation: New model from April, raised compression ratio, magneto, TT carb, big tank, siamezed pipes, frame as **Meteor Minor** but larger, casquette, air cleaner, clutch as **Meteor Minor**, steering damper. From September Airflow option available.

1959
Ensign III: No changes, discontinued mid-year.
Prince: New model as **III** with full flywheels, no decompressor, revised internals, cradle frame with swinging fork, teles, well valanced mudguards, extensive chaincase, full width hubs, side covers, dualseat, 3 gallon tank, Airflow option.
Clipper II: Dualseat standard, Airflow option.
Crusader: Alloy head, new silencer, Airflow option.
Crusader Sports: New model as **Crusader**, alloy head, Monobloc, internals changed, new silencer, 7 in. front brake, larger tank, footrests up and back, sports bars, Airflow option.
350 Clipper: Dualseat, chain enclosure and Airflow as options. Discontinued mid-year.
Bullets: **350** with revised internals, 17 in. wheels, 7 in. brakes, larger tank, chain enclosure and Airflow options. **500** with new cylinder head in alloy with cast-in rocker boxes, 6 in. dual front and 7 in. rear brakes, ignition switch in right toolbox with coil, Airflow option.
New **350 Trials** model with all alloy engine, heavy flywheels, magneto, inboard silencer.
Meteor Minor: No changes, standard model discontinued in year.
Super Meteor: Siamezed pipes.
Constellation: No changes.

1960
Prince: No changes.
Clipper II: Improved clutch.
Crusader & **Crusader Sports**: Improved clutch, new casquette.
350 Clipper: Alloy head, remainder as **Bullet**, 17 in. wheels, alternator, coil, new casquette.
Bullets: Coil ignition, new casquette. **500** had new tank and exhaust bend, 6 in. dual front brakes.
Meteor Minor: Improved clutch, new exhaust.
Meteor Minor Sports: New model with improved engine, larger tank.
Super Metoer & **Constellation**: 5 plate clutch, changed flywheel shape, oil breather in mainshaft, magnet in oil filter, new casquette style, twin Monoblocs on **Constellation**.

1961
Prince & **350 Clipper**: No changes.
Clipper II, **Crusader** & **Crusader Sports**: New silencer, minor change to casquette, sports exhaust and bigger tank for **Crusader**, tank only on Clipper.
Bullets: New silencer, 3-piece with fishtail and bottom outlet.
Meteor Minor: New silencer, restyled casquette.
Meteor Minor Sports: Engine parts satin finished, standard silencer, restyled casquette.
Super Meteor & **Constellation**: Partial rear enclosure with tail unit with seat, subframe and lifting handles, 3-part silencer, common fuel tank with modified mounting, siamesed pipes, lower bars on **Constellation**.

1962
Prince: Minor clutch mechanism change, discontinued in year.
Clipper II: No changes.
Crusader & **350 Clipper**: No changes, discontinued in year.
Crusader Sports: Styled rear mudguard as **Super 5**, 5 speed gearbox option in year.
Super 5: New sports model, tuned **Crusader** engine, 5 speed gearbox, standard frame, styled rear mudguard, leading link front forks, 7 in. front brake with floating anchor, sprung mudguard, new pressed steel casquette, no pilots, drop bars.

250 Trials: **Crusader Sports** engine, high level exhaust, special forks, trials equipment, coil ignition, alternator, battery, bulb horn.
Bullets: Deeper pressed steel rear mudguard, discontinued in year.
Meteor Minor: No changes, discontinued in year.
Meteor Minor Sports: Dual 6 in. front brakes, name changed late in year to **500 Sports Twin**.
Super Meteor & **Constellation**: No changes, discontinued in year.
Sidecar Constellation: New model from September with lower power engine, full sidecar specification of forks, gearing, tyres, rear units and steering damper; coil, single carb, siamezed pipes, two-pipe and Airflow options.
736 Interceptor: New model from September, enlarged **Constellation**, twin carbs, parallel mounted siamezed pipes, magneto, alternator, extra support bolted to rear of gearbox, flexible oil pipes to rockers, new prop stand, normal rear mudguard, Airflow option.

1963
Clipper II: Alloy head, bonded rubber rear fork pivot.
Crusader Sports: Normal rear mudguard, 5 speed option, new silencer, fork pivot as **Clipper**.
Super 5: Unsprung front mudguard, compression ratio as Sports, fork pivot as **Clipper**, discontinued in year.
250 Trials: Low gearing option, fork pivot as **Clipper**, on special order only in year.
Continental: New model with **Super 5** engine unit in standard frame with teles, fork pivot as **Clipper**, styled fuel tank, 7 in. front brake with air scoop, cast alloy fork crown, ball ended levers, dropped bars, exposed rear springs, flyscreen.
New Bullet: New model of 350 cc based on **Crusader** unit, revised internals, strengthened gearbox, fork pivot as **Clipper**.
Airflow available for **Clipper**, **Crusader Sports** and **Bullet**; Sportsflow for all singles.
500 Sports Twin: Siamezed pipes, separate pipe option, 7 in. brakes, discontinued in year.
Sidecar Constellation: Discontinued in year.
Interceptor: No changes.

1964
Clipper II: New silencer and tank, 5 speed and leading link options.
Crusader Sports: As **Clipper** in de-luxe form, standard model with painted tank and mudguards.
Continental: As **Crusader**, no rev-counter or flyscreen on standard model.
New Bullet: New silencer and tank.
Turbo Twin: New model with Villiers 4T engine in **Crusader** cycle parts.
Turbo Twin Sports: New model early in year with drop bars.
Interceptor: Available in de-luxe with 12 volt electrics or standard with 6 volts, both with sidecar specification and Airflow option, standard model discontinued in year.

1965
Clipper II & **Crusader Sports**: Modified silencer.
Continental: As **Clipper** plus fork gaiters added.
Turbo Twins & **New Bullet**: No changes.
Olympic: New model with 4 speed **Crusader** engine in standard frame with leading link front forks and **Super 5** casquette and styled rear mudguard.
Continental GT: Café racer style Continental, bellmouth, cranked exhaust, breather pipe, rearsets, clip-ons, alloy discs on front hub, exposed rear springs, gaitered forks, rev-counter, flyscreen, humped seat, scolloped tank, exposed battery and ignition switch.
Interceptor: Built in long wheelbase USA form only, separate headlamp, fascia panel.
In August *Continental*, **New Bullet**, **Turbo Twin** standard and **Olympic** discontinued.
In December **Clipper** and **Interceptor** in UK discontinued.

1966
Crusader Sports: No changes, discontinued in mid-year.
Turbo Twin Sports: No changes, discontinued late in year.
GT: Cover over coil and ignition switch.
Interceptor: Built for US only.

1967
GT: Discontinued in January.
Interceptor: On home market in September as in 1965 with cooling discs on front hub, coil ignition, 2 gallon tank, Concentrics, twin pipes.

1968
Interceptor: Continued, superseded in October by:
Series II: Wet sump, points in timing case, capacitor ignition, single oil pump in timing case, vertical oil filler, Norton forks, Norton 8 in. front brake, ignition switch on right panel, options of air cleaner, oil cooler, skid plate and seat rail. 2 gallon tank standard, 4 gallon option.

1969
Interceptor II: No changes.

1970
Interceptor II: Options fitted as standard except tank which was 4 gallon in UK and 2 in USA. Production stopped in mid-year.
Rickman Metisse: Series **II** engine in Rickman duplex frame, eccentric chain adjustors, seat unit, footrests mounted on exhaust pipes, disc brakes, alloy rims, slight step in seat. Became **Rickman Enfield**.

1971/72
Rickman Enfield: No changes to completion of build using surplus engines. 3 gallon fuel tank.

1977
Enfield India: **Bullet** built in India to 1955 specification in general but with ac electrics and coil ignition. Points in housing behind cylinder, casquette, dualseat with hand rail, number plate area on front mudguard sides, indicators.

1981/82
Enfield India: As in 1977 but no indicators.

Curious Pashley three-wheeler based on Enfield single. A style that was far more common in Italy using a Guzzi than in England

The Berkeley four-wheeler fitted with the Enfield engine. The resultant higher bonnet line was not as sleek as before

Model charts

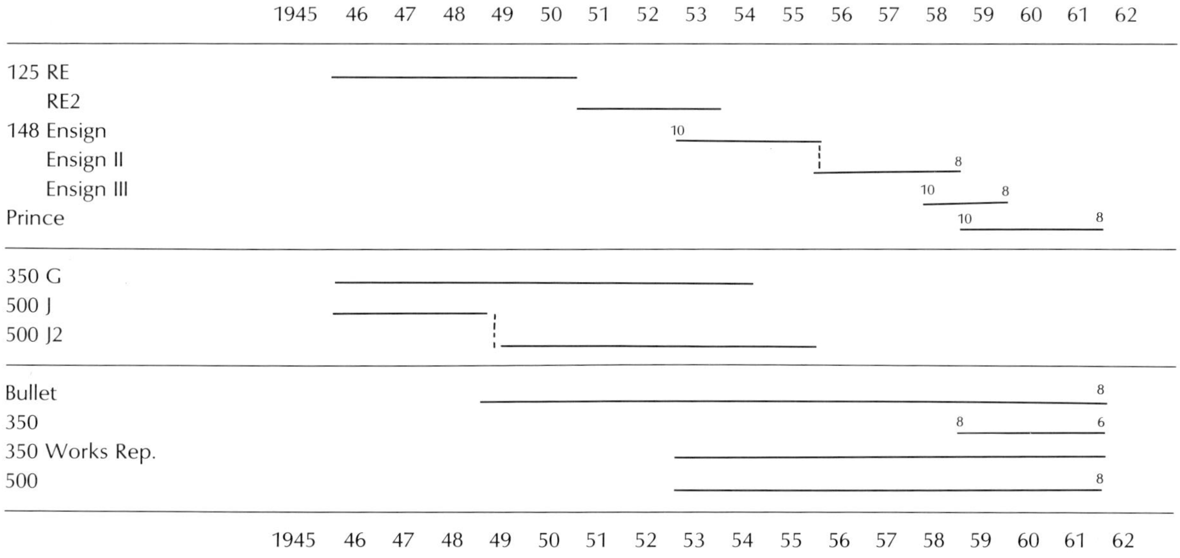

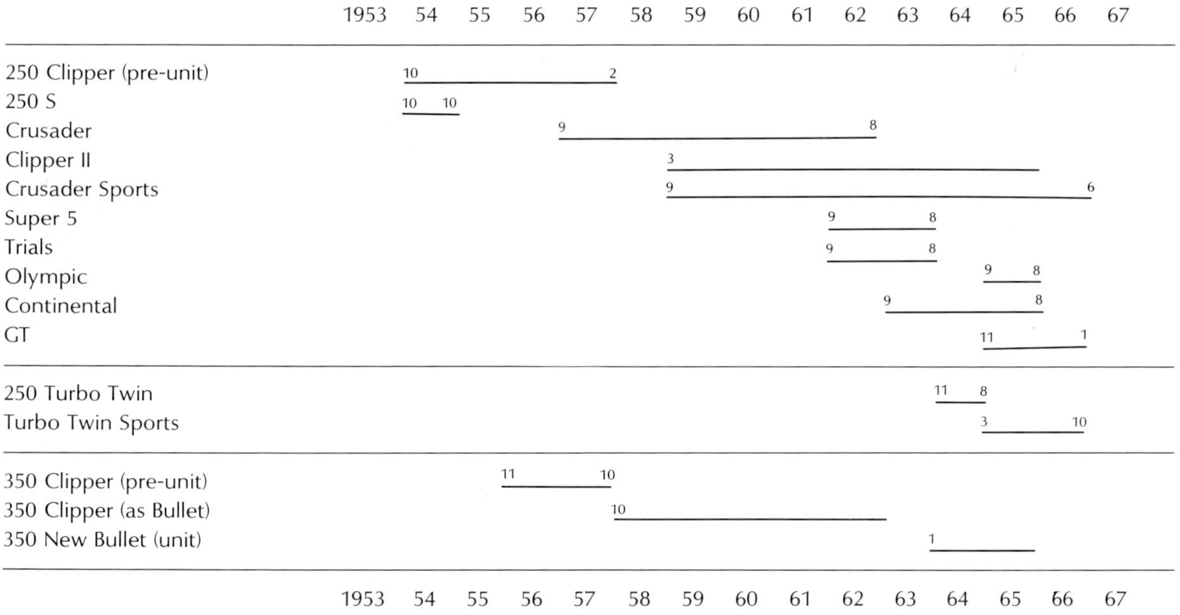